Stephen P. Sharples

Chemical Tables

Stephen P. Sharples

Chemical Tables

ISBN/EAN: 9783743329645

Manufactured in Europe, USA, Canada, Australia, Japa

Cover: Foto ©berggeist007 / pixelio.de

Manufactured and distributed by brebook publishing software
(www.brebook.com)

Stephen P. Sharples

Chemical Tables

CHEMICAL TABLES.

BY

STEPHEN P. SHARPLES, S.B.

CAMBRIDGE:

SEVER AND FRANCIS,

BOOKSELLERS TO THE UNIVERSITY.

1866.

3157

UNIVERSITY PRESS: WELCH, BIGELOW, & CO.,
CAMBRIDGE.

PREFACE.

THE following work was undertaken at the suggestion of Dr. Wolcott Gibbs, and has been executed under his immediate supervision. No labor has been spared to render the collection of tables as complete as possible, and to insure perfect accuracy in the figures. In all cases the tables have been taken from the original sources, or have been carefully compared with them. The tables on pages 94 – 99, 100 – 108, are new, and have been computed expressly for this work: the logarithms at the end of the volume are taken from the well-known five-figure table of August.

Many tables have been introduced, which are not in common use in the laboratory. This has been done partly for the sake of completeness, but principally because the progress of science continually brings the physical and chemical properties of bodies into closer relations with each other. Thus the researches of Landolt have recently shown that the chemical constitution of a mixture or compound may, in many cases, be deduced with accuracy from its optical properties alone.

The tables usually found in works on chemistry, giving the vapor densities of bodies referred to air taken as unity, have been omitted in this work, for the reason that it is always more convenient to refer gases and vapors to hydrogen as the unit of density as well as of weight and volume, and the vapor density is then very simply and directly related to the atomic weight.

The table of wave lengths on page 148 will be found convenient for use in connection with Kirchhoff's chart of the spectrum. With its assistance the wave length of any observed line may be determined with sufficient accuracy by the ordinary methods of interpolation.

The Editor will be greatly obliged for any criticisms or suggestions which may enable him to render the work more perfect in case it should pass to a second edition.

S. P. S.

LAWRENCE SCIENTIFIC SCHOOL,
Cambridge, June, 1866.

CONTENTS.

TABLES

FOR THE

CALCULATION OF ANALYSES.

TABLE

OF THE ELEMENTS AND THEIR EQUIVALENTS.

Element.	Symbol.	Equiva-lent.	Log.	Remarks.
Aluminum	Al	13.75	1.138303	
Antimony	Sb	120.00	2.079181	122 Dexter, 120.3 Schneider.
Arsenic	As	75.00	1.875061	
Barium	Ba	68.50	1.835691	68.59 Marignac.
Bismuth	Bi	208.00	2.318063	
Boron	B	11.00	1.041393	11.04 Berzelius.
Bromine	Br	80.00	1.903090	79.97 Marignac.
Cadmium	Cd	56.00	1.748188	
Cæsium	Cs	133.00	1.123852	
Calcium	Ca	20.00	1.301030	
Carbon	C	6.00	0.778151	
Cerium	Ce	46.00	1.662758	
Chlorine	Cl	35.50	1.550228	35.46 Marignac.
Chromium	Cr	26.00	1.414973	26.24 Berlin, Peligot.
Cobalt	Co	30.00	1.477121	
Copper	Cu	31.75	1.501744	31.73 Erdmann & Marchand.
Didymium	D	48.00	1.681241	
Erbium	E	56.30	1.750508	
Fluorine	F	19.00	1.278754	
Glucinum	G	4.66	0.668386	
Gold	Au	197.00	2.294466	196.67 Berzelius.
Hydrogen	H	1.00	0.000000	
Indium	In	37.00	1.568202	
Iodine	I	127.00	2.103804	
Iridium	Ir	98.00	1.991226	
Iron	Fe	28.00	1.447158	
Lanthanum	La	47.00	1.672098	46.4 Holzmann.
Lead	Pb	103.50	2.014940	103.57 Berzelius.
Lithium	L	7.00	0.845098	6.95 Mallet.
Magnesium	Mg	12.00	1.079181	[Hauer.
Manganese	Mn	27.00	1.431364	27.57 Berzelius, 27.5 Von
Mercury	Hg	100.00	2.000000	100.05 Erdmann & Marchand.
Molybdenum	Mo	46.00	1.662758	
Nickel	Ni	29.00	1.462398	
Niobium	Nb	94.00	1.973128	When NbO_5 = Niobic acid.
Nitrogen	N	14.00	1.146128	

Element.	Symbol.	Equiva- lent.	Log.	Remarks.
Osmium	Os	97.00	1.986772	99.82 Fremy.
Oxygen	O	8.00	0.903090	
Palladium	Pd	53.20	1.725912	53.24 Berzelius.
Phosphorus	P	31.00	1.491362	
Platinum	Pt	99.00	1.995635	98.94 Andrews.
Potassium	K	39.00	1.591065	39.11 Marignac.
Rhodium	R	52.00	1.716003	52.16 Berzelius.
Rubidium	Rb	85.40	1.931458	
Ruthenium	Ru	52.00	1.716003	
Selenium	Se	39.50	1.596597	39.41 Erdmann & Marchand.
Silicon	Si	14.00	1.146128	When Silicic acid $= SiO_2$.
Silver	Ag	108.00	2.033424	
Sodium	Na	23.00	1.361728	[louze.
Strontium	Sr	43.75	1.640978	43.67 Stromeyer, 43.85 Pe-
Sulphur	S	16.00	1.204120	
Tantalum	Ta	172.00	2.235528	When Tantalic acid $= TaO_5$.
Tellurium	Te	64.00	1.806180	64.14 Berzelius.
Thallium	Tl	204.00	2.309630	
Thorium	Th	118.00	2.071882	When Thoria $= ThO_2$.
Tin	Sn	58.00	1.763428	
Titanium	Ti	25.00	1.397940	
Tungsten	W	92.00	1.963788	
Uranium	U	60.00	1.778151	
Vanadium	V	68.50	1.835691	68.54 Berzelius.
Yttrium	Y	30.85	1.489255	
Zinc	Zn	32.50	1.511883	32.52 Axel, Erdmann.
Zirconium	Zr	44.80	1.651278	When Zirconia $= ZrO_2$.

TABLE

GIVING THE ELEMENTS ARRANGED IN NATURAL GROUPS; WITH
THEIR MOLECULAR AND ATOMIC WEIGHTS.

1. *Monatomic Elements, or Elements usually equivalent to 1 Atom of Hydrogen.*

Element.	Atomic Weight.	Molecular Weight.	Element.	Atomic Weight.	Molecular Weight.
Bromine	80.00	160.00	Lithium	7.00	14.00
Chlorine	35.50	71.00	Potassium	39.00	78.00
Cæsium	133.00	266.00	Rubidium	85.40	170.80
Fluorine	19.00	38.00	Silver	108.00	216.00
Gold	197.00	394.00	Sodium	23.00	46.00
Hydrogen	1.00	2.00	Thallium	204.00	408.00
Iodine	127.00	254.00			

2. *Diatomic Elements, or Elements usually equivalent to 2 Atoms of Hydrogen.*

Element.	Atomic Weight.	Molecular Weight.	Element.	Atomic Weight.	Molecular Weight.
Oxygen	16.00	32.00	Indium?	74.00	148.00
Sulphur	32.00	64.00	Cobalt	60.00	120.00
Selenium	79.00	158.00	Nickel	58.00	116.00
Tellurium	128.00	256.00	Cadmium	112.00	224.00
Calcium	40.00	80.00	Copper	63.50	127.00
Barium	137.00	274.00	Mercury	200.00	400.00
Strontium	87.50	175.00	Cerium	92.00	184.00
Lead	207.00	414.00	Didymium	96.00	192.00
Iron	56.00	112.00	Lanthanum	94.00	188.00
Manganese	54.00	108.00	Yttrium	61.70	123.40
Chromium	52.00	104.00	Erbium	112.60	225.20
Aluminum	27.50	55.00	Glucinum	9.32	18.64
Zinc	65.00	130.00	Uranium	120.00	240.00
Magnesium	24.00	48.00			

3. *Triatomic Elements, or Elements usually equivalent to 3 Atoms of Hydrogen.*

Element.	Atomic Weight.	Molecular Weight.	Element.	Atomic Weight.	Molecular Weight.
Nitrogen	14.00	28.00	Boron	11.00	?
Phosphorus	31.00	124.00	Molybdenum?	92.00	?
Arsenic	75.00	300.00	Tungsten?	184.00	?
Antimony	120.00	480.00	Vanadium?	137.00	?
Bismuth	208.00	832.00			

4. *Tetratomic Elements, or Elements usually equivalent to 4 Atoms of Hydrogen.*

Element.	Atomic Weight.	Molecular Weight.	Element.	Atomic Weight.	Molecular Weight.
Carbon	12.00	24.00 ?	Osmium	194.00	388.00 ?
Silicon	28.00	56.00 ?	Palladium	106.40	212.80 ?
Zirconium	89.60	179.20 ?	Platinum	198.00	396.00 ?
Thorium	236.00	472.00 ?	Iridium	196.00	392.00 ?
Tin	116.00	232.00 ?	Rhodium	104.00	208.00 ?
Titanium	50.00	100.00 ?	Ruthenium	104.00	208.00 ?

5. *Pentatomic Elements, or Elements usually equivalent to 5 Atoms of Hydrogen.*

Element.	Atomic Weight.	Molecular Weight.	Element.	Atomic Weight.	Molecular Weight.
Tantalum	344.00	688.00 ?	Niobium	198.00	396.00 ?

TABLE

OF THE FORMULAS, EQUIVALENTS, AND PER CENTS OF THE CONSTITUENTS, OF THE MOST FREQUENTLY OCCURRING COMPOUNDS.

Combination	Formula	Equivalent.	Per cent of Constituents.
Bromides.			
Bromide of Potassium	KBr	119.00	K 32.77 Br 67.23
" Silver	AgBr	188.00	Ag 57.45 Br 42.55
" Sodium	NaBr	103.00	Na 22.33 Br 77.67
Chlorides.			
Chloride of Ammonium	NH$_4$Cl	53.50	NH$_4$ 33.65 Cl 66.35
" Ammonium & Platinum	NH$_4$Cl, PtCl$_2$	223.50	NH$_3$ 31.78 HCl 68.22
			NH$_4$ 8.06 Pt 44.29 Cl 47.65 ; or,
			NH$_4$Cl 23.94 PtCl$_2$ 76.06 ; or,
			NH$_3$ 7.60 HCl 16.34 PtCl$_2$ 76.06
			N 6.26 H 1.79 Cl 47.65 Pt 44.30
" Barium	BaCl	104.00	Ba 65.86 Cl 34.14
" Crystal.	BaCl + 2HO	122.00	Ba 56.15 Cl 29.09 HO 14.76
" Calcium	CaCl	55.50	Ca 36.04 Cl 63.96
" Cobalt	CoCl	65.50	Co 45.80 Cl 54.20
" Crystal.	CoCl + 6HO	119.50	Co 25.10 Cl 29.71 HO 45.19
" Copper	CuCl	67.25	Cu 47.21 Cl 52.79
" Sub	Cu$_2$Cl	99.00	Cu 64.15 Cl 35.85
" Gold	AuCl$_3$	303.50	Au 64.91 Cl 35.09
" Iron Proto	FeCl	63.50	Fe 44.10 Cl 55.90
" Sesqui	Fe$_2$Cl$_3$	162.50	Fe 34.46 Cl 65.54

Combination.	Formula.	Equivalent.	Per cent of Constituents.
Chloride of Lead	PbCl	139.00	Pb 74.46 Cl 25.54
" Magnesium	MgCl	47.50	Mg 25.26 Cl 74.74
" Manganese	MnCl	62.50	Mn 43.20 Cl 56.80
" " Crystal.	$MnCl + 4HO$	98.50	Mn 27.41 Cl 36.04 HO 36.55
" Mercury	HgCl	135.50	Hg 73.80 Cl 26.20
" " Sub	Hg_2Cl	235.50	Hg 84.92 Cl 15.08
" Platinum	PtCl	134.50	Pt 73.61 Cl 26.39
" " Bi	$PtCl_2$	170.00	Pt 58.23 Cl 41.77
" Potassium	KCl	74.50	K 52.35 Cl 47.65
" " & Platinum	$KCl, PtCl_2$	244.50	{ KCl 30.47 $PtCl_2$ 69.53 ; or, K 15.95 Pt 40.49 Cl 43.56
" Silver	AgCl	143.50	Ag 75.26 Cl 24.74
" Sodium	NaCl	58.50	Na 39.32 Cl 60.68
" Tin Bi	$SnCl_2$	129.00	Sn 44.96 Cl 55.04
" " Crystal.	$SnCl_2 + 2HO$	111.50	Sn 52.02 Cl 31.84 HO 16.14
" " Proto	SnCl	93.50	Sn 62.03 Cl 37.97
" Zinc	ZnCl	68.00	Zn 47.79 Cl 52.21

Cyanides.

Combination.	Formula.	Equivalent.	Per cent of Constituents.
Cyanogen = Cy	C_2N	26.00	C 46.15 N 53.85
Cyanide of Mercury	HgCy	126.00	Hg 79.36 Cy 20.64
" Palladium	PdCy	79.20	Pd 67.18 Cy 32.82
" Potassium	KCy	65.00	K 60.00 Cy 40.00
" Silver	AgCy	134.00	Ag 80.59 Cy 19.41
Ferricyanide of Potassium	$3KCy, Fe_2Cy_3$	329.00	KCy 59.27 Fe_2Cy_3 40.73
Ferrocyanide of Potassium	$2KCy, FeCy + 3HO$	211.00	KCy 61.61 Fe Cy 25.59 HO 12.80
Sulphocyanide of Potassium	$KCyS_2$	97.00	K 40.20 Cy 26.81 S 32.99

Fluorides.

Fluoride of Ammonium	NH_4F	37.00	NH_4 48.65 F 51.35
" Calcium	CaF	39.00	Ca 51.28 F 48.72
" Sodium & Aluminum	$3NaF, Al_2F_3$	210.50	Na 32.77 Al 13.06 F 54.17
" Potassium	KF	58.00	K 67.24 F 32.76
" Sodium	NaF	42.00	Na 54.76 F 45.24
Fluosilicate of Barium	BaF, BF_3	139.50	Ba 49.10 Si 10.04 F 40.86
" Potassium	KF, SiF_2	110.00	K 35.45 Si 12.73 F 51.82
Fluoborate of "	KF, BF_3	126.00	K 30.95 B 8.73 F 60.32

Compounds of Hydrogen.

Hydrobromic acid	HBr	81.00	H 1.24 Br 98.76
Hydrochloric "	HCl	36.50	H 2.74 Cl 97.26
Hydrocyanic "	HCy	27.00	H 3.71 Cy 96.29
Hydrofluoric "	HF	20.00	H 5.00 F 95.00
Hydroiodic "	HI	128.00	H 0.78 I 99.22
Hydrosulphuric acid	HS	17.00	H 5.88 S 94.12
Ammonia	NH_3	17.00	H 17.65 N 82.35

Iodides.

Iodide of Copper (Sub)	Cu_2I	190.50	Cu 33.34 I 66.66
" Mercury	HgI	227.00	Hg 44.06 I 55.94
" Palladium	PdI	180.20	Pd 29.58 I 70.42
" Potassium	KI	166.00	K 23.48 I 76.52
" Silver	AgI	235.00	Ag 45.96 I 54.04
" Sodium	NaI	150.00	Na 15.34 I 84.66

Oxides.

Aluminum, Oxide	Al_2O_3	51.50	Al 53.40 O 46.60
Ammonium, Oxide	NH_4O	26.00	NH_4 69.23 O 30.77
Antimony, Oxide	SbO_4	144.00	Sb 83.35 O 16.65

Combination.	Formula.	Equiva-lent.	Per cent of Constituents.
Antimonous Acid	SbO_4	152.00	Sb 78.96 O 21.04
Antimonic Acid	SbO_5	160.00	Sb 75.01 O 24.99
Arsenous Acid	AsO_3	99.00	As 75.76 O 24.24
Arsenic Acid	AsO_5	115.00	As 65.22 O 34.78
Barium, Oxide	BaO	76.50	Ba 89.55 O 10.45
" Hydrate	BaO, HO	85.50	BaO 89.48 HO 10.52
" Peroxide	BaO_2	84.50	Ba 81.06 O 18.94
Bismuth, Oxide	BiO_3	232.00	Bi 89.65 O 10.35
Boric Acid	BO_3	35.00	B 31.43 O 68.57
" " Crystal.	$BO_3 + 3HO$	62.00	BO 56.44 HO 43.56
Cadmium, Oxide	CdO	64.00	Cd 87.50 O 12.50
Calcium "	CaO	28.00	Ca 71.43 O 28.57
" Hydrate	CaO, HO	37.00	CaO 75.67 HO 24.33
Carbonic Oxide	CO	14.00	C 42.86 O 57.14
" Acid	CO_2	22.00	C 27.27 O 72.73
Chloric Acid	ClO_5	75.50	Cl 47.02 O 52.98
Chromic Acid	CrO_3	50.00	Cr 52.00 O 48.00
Chromium, Sesquioxide	Cr_2O_3	76.00	Cr 68.42 O 31.58
Cobalt, Oxide	CoO	38.00	Co 78.94 O 21.06
" Sesquioxide	Co_2O_3	84.00	Co 71.43 O 28.57
Cobalt, Protosesquioxide	Co_3O_4	122.00	Co 73.77 O 26.23
Copper, Oxide	CuO	39.75	Cu 79.88 O 20.12
" Suboxide	Cu_2O	71.50	Cu 88.81 O 11.19
Gold, Teroxide	AuO	221.00	Au 89.14 O 10.86
Hydrogen, Oxide	HO	9.00	H 11.11 O 88.89
" Peroxide	HO_2	17.00	H 5.88 O 94.12
Iodic Acid	IO_5	167.00	I 76.05 O 23.95
Iron, Oxide	FeO	36.00	Fe 77.78 O 22.22

Iron, Sesquioxide	Fe_2O_3	80.00	Fe 70.00	O 30.00		
" Magnetic Oxide	Fe_3O_4	116.00	Fe 72.41	O 27.59		
Lead, Suboxide	Pb_2O	215.00	Pb 96.28	O 3.72		
" Oxide	PbO	111.50	Pb 92.83	O 7.17		
" Peroxide	PbO_2	119.50	Pb 86.61	O 13.39		
Magnesium, Oxide	MgO	20.00	Mg 60.00	O 40.00		
" Hydrate	MgO, HO	29.00	MgO 68.97	HO 31.03		
Manganese, Oxide	MnO	35.00	Mn 77.14	O 22.86		
" Sesquioxide	Mn_2O_3	78.00	Mn 69.23	O 30.77		
" Protosesquioxide	Mn_3O_4	113.00	Mn 71.68	O 28.32		
" Peroxide	MnO_2	43.00	Mn 62.79	O 37.21		
Manganic Acid	MnO_3	51.00	Mn 52.94	O 47.06		
Permanganic Acid	Mn_2O_7	110.00	Mn 49.09	O 50.91		
Mercury, Oxide	HgO	108.00	Hg 92.59	O 7.41		
" Suboxide	Hg_2O	208.00	Hg 96.15	O 3.85		
Molybdic Acid	MoO_3	70.00	Mo 65.71	O 34.29		
Nickel, Oxide	NiO	37.00	Ni 78.38	O 21.62		
Nitrogen, Oxide	NO	22.00	N 63.64	O 36.36		
" Deutoxide	NO_2	90.00	N 46.67	O 53.33		
Nitrous Acid	NO_3	38.00	N 36.84	O 63.16		
Nitric Acid	NO_5	54.00	N 25.93	O 74.07		
" Hydrate	HO, NO_5	63.00	NO_5 85.71	HO 14.29		
Oxalic Acid	C_2O_3	36.00	C 33.33	O 66.67		
" " Crystal.	$C_2O_3 + 3HO$	63.00	$\bar{O}$ 57.14	HO 42.86		
Phosphoric Acid	PO_5	71.00	P 43.66	O 56.34		
" " Hydrated	HO, PO_5	80.00	HO 11.25	PO_5 88.75		
Pyrophosphoric Acid	$2HO, PO_5$	89.00	HO 20.23	PO_5 79.77		
Phosphoric Acid, common	$3HO, PO_5$	98.00	HO 27.55	PO_5 72.45		
Platinum, Binoxide	PtO_2	115.00	Pt 86.08	O 13.92		
Potassium, Oxide	KO	47.00	K 82.94	O 17.06		

Combination.	Formula.	Equivalent.	Per cent of Constituents.
Potassium, Hydrate	KO, HO	56.00	KO 83.93 HO 16.07
Silicic Acid	SiO_3	30.00	Si 46.67 O 53.33
Silver, Oxide	AgO	116.00	Ag 93.10 O 6.90
Sodium, Oxide	NaO	31.00	Na 74.19 O 25.81
" Hydrate	NaO, HO	40.00	Na 77.50 HO 22.50
Strontium, Oxide	SrO	51.75	Sr 84.54 O 15.46
Sulphuric Acid	SO_3	40.00	S 40.00 O 60.00
" " Hydrate	HO, SO_3	49.00	SO_3 81.63 O 18.37
Sulphurous Acid	SO_2	32.00	S 50.00 O 50.00
Stannic Acid	SnO_2	74.00	Sn 78.38 O 21.62
Tin, Oxide	SnO	66.00	Sn 87.88 O 12.12
Titanic Acid	TiO_2	41.00	Ti 60.98 O 39.02
Uranium, Oxide	UO_2	68.00	U 88.24 O 11.76
" Sesquioxide	U_2O_3	144.00	U 83.33 O 16.67
" Protosesquioxide	U_2O_3	212.00	U 84.91 O 15.09
Zinc, Oxide	ZnO	40.50	Zn 80.24 O 19.76

Sulphides.

Combination.	Formula.	Equivalent.	Per cent of Constituents.
Ammonium, Sulphide	NH_4S	34.00	NH_4 52.94 S 47.06
Antimony, Tersulphide	SbS_3	168.00	Sb 71.43 S 28.57
" Pentasulphide	SbS_5	200.00	Sb 60.00 S 40.00
Arsenic, Bisulphide	AsS_2	107.00	As 70.09 S 29.91
" Tersulphide	AsS_3	123.00	As 60.98 S 39.02
" Pentasulphide	AsS_5	155.00	As 48.39 S 51.61
Cadmium, Sulphide	CdS	72.00	Cd 77.78 S 22.22
Carbon, Bisulphide	CS_2	38.00	C 15.79 S 84.21
Copper, Sulphide	CuS	47.75	Cu 66.49 S 33.51

Name	Formula	Wt.	Composition
Copper, Subsulphide	Cu₂S	79.50	Cu 79.87 S 20.13
Iron, Sulphide	FeS	44.00	Fe 63.64 S 36.36
" Bisulphide	FeS₂	60.00	Fe 46.67 S 53.33
Lead, Sulphide	PbS	119.50	Pb 86.61 S 13.39
Mercury, Sulphide	HgS	116.00	Hg 86.21 S 13.79
Molybdenum, Sulphide	MoS₂	78.00	Mo 58.97 S 41.03
Potassium, "	KS	55.00	K 70.91 S 29.09
" Tersulphide	KS₃	87.00	K 44.83 S 55.17
" Pentasulphide	KS₅	119.00	K 32.77 S 67.23
Silver, Sulphide	AgS	124.00	Ag 87.09 S 12.91
Sodium, "	NaS	39.00	Na 58.97 S 41.03
" Tersulphide	NaS₃	71.00	Na 32.39 S 67.61
" Pentasulphide	NaS₅	103.00	Na 22.33 S 77.67
Tin, Sulphide	SnS	74.00	Sn 78.38 S 21.62
" Bisulphide	SnS₂	90.00	Sn 64.44 S 35.56
Zinc, Sulphide	ZnS	48.50	Zn 67.01 S 32.99

Acetates.

$$\overline{A} = C_4H_3O_3 = 51.$$

Base.		Wt.	Composition
Copper	CuO, $\overline{A}$ + HO	99.75	CuO 39.85 $\overline{A}$ 51.13 HO 9.02
Lead	PbO, $\overline{A}$ + 3HO	189.50	PbO 58.84 $\overline{A}$ 26.91 HO 14.25
Potassium	KO, $\overline{A}$	98.00	KO 47.99 $\overline{A}$ 52.01
Silver	AgO, $\overline{A}$	167.00	Ag 69.46 $\overline{A}$ 30.54
Sodium	NaO, $\overline{A}$ + 6HO	136.00	NaO 22.79 $\overline{A}$ 37.50 HO 39.71

Borates.

		Wt.	Composition
Sodium, Bi-	NaO, 2BoO₃	101.00	NaO 30.70 BO₃ 69.30
" Crystalized	NaO, 2BoO₃ + 10HO	191.00	NaO 16.23 BO₃ 36.65 HO 47.12

Carbonates.

Base.	Formula.	Equivalent.	Per cent of Constituents.
Ammonium, Sesqui	$2NH_4O,3CO_2$	118.00	NH_4O 44.07 CO_2 55.93
" Neutral	NH_4O,CO_2	48.00	NH_4O 54.17 CO_2 45.83
Barium	BaO,CO_2	98.50	BaO 77.66 CO_2 22.34
Calcium	CaO,CO_2	50.00	CaO 56.00 CO_2 44.00
Iron	FeO,CO_2	58.00	FeO 62.07 CO_2 37.93
Magnesium	MgO,CO_2	42.00	MgO 47.62 CO_2 52.38
Manganese	MnO,CO_2	57.00	MnO 61.40 CO_2 38.60
Lead	PbO,CO_2	133.50	PbO 83.52 CO_2 16.48
Potassium	KO,CO_2	69.00	KO 68.12 CO_2 31.88
" Bi	$KO,2CO_2 + HO$	100.00	KO 47.00 CO_2 44.00 HO 9.00
Sodium	NaO,CO_2	53.00	NaO 58.49 CO_2 41.51
" Crystal.	$NaO,CO_2 +10HO$	143.00	NaO 21.67 CO_2 15.39 HO 62.94
" Bi	$NaO,2CO_2 + HO$	84.00	NaO 36.90 CO_2 52.38 HO 10.72
Strontium	SrO,CO_2	73.75	SrO 70.15 CO_2 29.85
Zinc	ZnO,CO_2	62.50	ZnO 64.80 CO_2 35.20

Chlorates.

Base.	Formula.	Equivalent.	Per cent of Constituents.
Barium	BaO,ClO_5	152.00	BaO 50.33 ClO_5 49.67
Potassium	KO,ClO_5	122.50	KO 38.37 ClO_5 61.63
Silver	AgO,ClO_5	191.50	AgO 60.57 ClO_5 39.43
Sodium	NaO,ClO_5	106.50	NaO 29.11 ClO_5 70.89

Chromates.

Base.	Formula.	Equivalent.	Per cent of Constituents.
Barium	BaO,CrO_3	126.50	BaO 60.47 CrO_3 39.53
Bismuth	$BiO_3,2CrO_3$	332.00	BiO_3 69.88 CrO_3 30.12

Lead, Neutral	PbO, CrO_3	161.50	PbO 69.04 CrO_3 30.96
" Basic	$2PbO, CrO_3$	273.00	PbO 81.69 CrO_3 18.31
Mercury	$4Hg_2O, 3CrO_3$	982.00	Hg_2O 84.72 CrO_3 15.28
Potassium, Neutral	KO, CrO_3	97.00	KO 48.45 CrO_3 51.55
" Acid	$KO, 2CrO_3$	147.00	KO 31.96 CrO_3 68.04
Silver, Acid	$AgO, 2CrO_3$	216.00	AgO 53.70 CrO_3 46.30

Nitrates.

Ammonium	NH_4O, NO_6	80.00	NH_4O 32.50 NO_5 67.50
Barium	BaO, NO_6	130.50	BaO 58.62 NO_6 41.38
Bismuth	$BiO_3, 3NO_5 + 10HO$	484.00	BiO 47.94 NO_6 33.47 HO 18.59
Cobalt	$CoO, NO_6 + 6HO$	146.00	CoO 26.02 NO_6 36.99 HO 36.99
Calcium	CaO, NO_6	82.00	CaO 34.15 NO_6 65.85
Lead	PbO, NO_5	165.50	PbO 67.37 NO_6 32.63
Magnesium	MgO, NO_6	74.00	MgO 27.03 NO_6 72.97
Mercury, Sub	$Hg_2O, NO_5 + 2HO$	280.00	Hg_2O 74.29 NO_5 19.28 HO 6.43
Potassium	KO, NO_6	101.00	KO 46.54 NO_6 53.46
Silver	AgO, NO_6	170.00	AgO 68.23 NO_5 31.77
Sodium	NaO, NO_6	85.00	NaO 36.47 NO_6 63.53
Strontium	SrO, NO_6	105.75	SrO 48.93 NO_6 51.07

Oxalates.

$$\bar{O} = C_2O_3 = 36.$$

Calcium	$CaO, \bar{O} + HO$	73.00	CaO 38.35 $\bar{O}$ 49.32 HO 12.33
Potassium	$KO, \bar{O} + HO$	92.00	KO 51.09 $\bar{O}$ 39.12 HO 9.79
" Bin	$KO, 2\bar{O} + 3HO$	146.00	KO 32.19 $\bar{O}$ 49.31 HO 18.50
" Per	$KO, 4\bar{O} + 7HO$	254.00	KO 18.51 $\bar{O}$ 56.69 KO 24.80

Phosphates.

Base.	Formula.	Equivalent.	Per cent of Constituents.
Ammonium	$2NH_4O, HO, PO_5$	132.00	NH_4O 39.39 HO 6.82 PO_5 53.79
" and Magnesium	$NH_4O, 2MgO, PO_5 + 12HO$	245.00	NH_4O 10.61 MgO 16.33 PO_5 28.98 HO 44.08
Calcium	$3CaO, PO_5$	155.00	CaO 54.19 PO_5 45.81
Iron	Fe_2O_3, PO_5	151.00	Fe_2O_3 52.98 PO_5 47.02
Magnesium, Pyro	$2MgO, PO_5$	111.00	MgO 36.04 PO_5 63.96
Manganese, "	$2MnO, PO_5$	141.00	MnO 49.65 PO_5 50.35
Sodium	$2NaO, HO, PO_5 + 24HO$	358.00	NaO 17.32 PO_5 19.83 HO 62.85
Silver	$3AgO, PO_5$	419.00	AgO 83.05 PO_5 16.95
" Pyro	$2AgO, PO_5$	303.00	AgO 76.57 PO_5 23.43
Uranium	U_2O_3, PO_5	359.00	U_2O_3 80.22 PO_5 19.78

Sulphates.

Base.	Formula.	Equivalent.	Per cent of Constituents.
Aluminum	$Al_2O_3, 3SO_3 + 18HO$	333.50	Al_2O_3 15.44 SO_3 35.98 HO 48.58
Ammonium	NH_4O, SO_3	66.00	NH_4O 39.40 SO_3 60.60
" and Aluminum	$NH_4O, SO_3, Al_2O_3, 3SO_3 + 24HO$	453.50	NH_4O 5.73 Al_2O_3 11.36 SO_3 35.28 HO 47.63
Barium	BaO, SO_3	116.50	BaO 65.67 SO_3 34.33
Calcium	CaO, SO_3	68.00	CaO 41.18 SO_3 58.82
" Crystal.	$CaO, SO_3 + 2HO$	86.00	CaO 32.56 SO_3 46.51 HO 20.93
Cobalt	CoO, SO_3	78.00	CoO 48.72 SO_3 51.28
" Crystal.	$CoO, SO_3 + 7HO$	141.00	CoO 26.95 SO_3 28.37 HO 44.68
Copper	CuO, SO_3	79.75	CuO 49.84 SO_3 50.16
" Crystal.	$CuO, SO_3 + 5HO$	124.75	CuO 31.86 SO_3 32.07 HO 36.07

Iron	FeO,SO₃	76.00	FeO 47.37 SO₃ 52.63
" Crystal.	FeO,SO₃ + 7HO	139.00	FeO 25.89 SO₃ 28.77 HO 45.34
Magnesium	MgO,SO₃	60.00	MgO 33.33 SO₃ 66.67
" Crystal.	MgO,SO₃ + 7HO	123.00	MgO 16.26 SO₃ 32.52 HO 51.22
Manganese	MnO,SO₃	75.00	MnO 46.67 SO₃ 53.33
Nickel, Crystal.	NiO,SO₃ + 7HO	140.00	NiO 26.43 SO₃ 28.57 HO 45.00
Potassium	KO,SO₃	87.00	KO 54.02 SO₃ 45.98
" Bi	KO,2SO₃	127.00	KO 37.00 SO₃ 63.00
" and Aluminum	KO,SO₃,Al₂O₃ 3SO₃ + 24HO	474.50	{ KO 9.90 Al₂O₃ 10.86 / SO₃ 33.71 HO 45.52
Lead	PbO,SO₃	151.50	PbO 73.60 SO₃ 26.40
Sodium	NaO,SO₃	71.00	NaO 43.66 SO₃ 56.34
" Crystal.	NaO,SO₃ + 10HO	161 00	NaO 19.25 SO₃ 24.85 HO 55.90
Silver	AgO,SO₃	156.00	AgO 74.36 SO₃ 25.64
Strontium	SrO,SO₃	91.75	SrO 56.40 SO₃ 43.60
Zinc	ZnO,SO₃	80.50	ZnO 50.31 SO₃ 49.69
" Crystal.	ZnO,SO₃ + 7HO	143.50	ZnO 28.22 SO₃ 27.88 HO 43.90

Tartrates.

$$\bar{T} = C_8H_4O_{10} = 192.$$

Calcium	2CaO,T̄ + 8HO	260.00	CaO 21.54 T̄ 50.77 HO 27.69
Potassium	2KO,T̄ + HO	235.00	KO 40.00 T̄ 56.17 HO 3.83
" Acid	KO,HO,T̄	188.00	KO 25.00 T̄ 70.21 HO 4.79
" and Sodium	KO,NaO,T̄ + 8HO	282.00	{ KO 16.66 NaO 10.99. / T̄ 46.81 HO 25.54
" " Antimony	KO,SbO₂,T̄ + HO	332.00	{ SbO₂ 43.37 KO 14.16 / T̄ 39.76 HO 2.71

TABLE

OF FACTORS FOR THE REDUCTION OF THE SUBSTANCE FOUND
IN ANALYSIS TO THE ONE SOUGHT.

The weight found multiplied by the factor gives the weight of the substance sought.

FRESENIUS'S ANALYSIS.

Element.	Found.	Sought.	Factor.
Aluminum	Al_2O_3	Al	0.5340
Ammonium	NH_4Cl	NH_3	0.3178
	NH_4Cl	NH_4	0.3365
	$NH_4Cl, PtCl_2$	NH_3	0.0760
	$NH_4Cl, PtCl_2$	NH_4O	0.1163
Antimony	SbO_3	Sb	0.8335
	SbS_3	Sb	0.7143
	SbO_4	Sb	0.7896
	SbO_4	SbO_3	0.9474
Arsenic	AsO_3	As	0.7576
	AsO_5	As	0.6522
	AsO_5	AsO_3	0.8609
	AsS_3	As	0.6098
	AsS_3	AsO_3	0.8049
	AsS_3	AsO_5	0.9349
	$2MgO, NH_4O, AsO_5 + HO$	AsO_5	0.6053
	$2MgO, NH_4O, AsO_5 + HO$	AsO_3	0.5211
Barium	BaO	Ba	0.8955
	BaO, SO_3	BaO	0.6567
	$BaFl, SiFl_2$	BaO	0.5484
Bismuth	BiO_3	Bi	0.8965
Boron	BO_3	B	0.3143
Bromine	$AgBr$	Br	0.4255
Cadmium	CdO	Cd	0.8750
Cæsium	$CsCl$	Cs	0.7893
	CsO, SO_3	CsO	0.7790
Calcium	CaO	Ca	0.7143
	CaO, SO_3	CaO	0.4118
	CaO, CO_2	CaO	0.5600
Carbon	CO_2	C	0.2727
	CaO, CO_2	CO_2	0.4400
Cerium	Ce_3O_4	Ce	0.8118
Chlorine	$AgCl$	Cl	0.2474
	$AgCl$	HCl	0.2537
Chromium	Cr_2O_3	Cr	0.6842

Element.	Found.	Sought.	Factor.
Chromium	Cr_2O_3	CrO_3	0.6579
	PbO, CrO_3	CrO_3	0.3096
Cobalt	Co	CoO	1.2667
	CoO, SO_3	CoO	0.4872
Copper	CuO	Cu	0.7987
	Cu_2S	Cu	0.7987
Didymium	DO	D	0.8571
Erbium	ErO	Er	0.8756
Fluorine	CaF	F	0.4872
	SiF_2	F	0.7308
Glucinum	GO	G	0.3681
Hydrogen	HO	H	0.1111
Indium	InS	In	0.6981
Iodine	AgI	I	0.5404
	PdI	I	0.7042
Iron	Fe_2O_3	Fe	0.7000
	Fe_2O_3	FeO	0.9000
Lanthanum	LaO	La	0.8545
Lead	PbO	Pb	0.9283
	PbO, SO_3	Pb	0.6832
	$PbCl$	Pb	0.7446
	PbS	Pb	0.8661
Lithium	$3LO, PO_5$	LO	0.3879
	LCl	L	0.1647
Magnesium	MgO	Mg	0.6000
	MgO, SO_3	MgO	0.3333
	$2MgO, PO_5$	MgO	0.3604
Manganese	MnO	Mn	0.7714
	Mn_2O_3	MnO	0.8973
	Mn_3O_4	MnO	0.9292
	MnO, SO_3	MnO	0.4667
	$2MnO, PO_5$	MnO	0.4965
Mercury	Hg	Hg_2O	1.0400
	Hg	HgO	1.0800
	Hg_2Cl	Hg	0.8492
	HgS	Hg	0.8621
Molybdenum	MoO_3	Mo	0.6571
Nickel	NiO	Ni	0.7838
Niobium	NbO_5	Nb	0.7015
Nitrogen	$NH_4Cl, PtCl_2$	N	0.0627
	Pt	N	0.1414
	AgC_2N	C_2N	0.1941
	AgC_2N	HC_2N	0.2015
Oxygen	Al_2O_3	O	0.4660
	SbO_3	O	0.1665
	AsO_3	O	0.2424
	AsO_5	O	0.3478

Element.	Found.	Sought.	Factor.
Oxygen	BaO	O	0.1045
	BiO_3	O	0.1035
	CdO	O	0.1250
	CaO	O	0.2857
	Cr_2O_3	O	0.3158
	CuO	O	0.2012
	HO	O	0.8889
	Fe_2O_3	O	0.3000
	FeO	O	0.2222
	PbO	O	0.0717
	MgO	O	0.4000
	MnO	O	0.2286
	Mn_3O_4	O	0.2832
	Mn_2O_3	O	0.3077
	MnO_2	O	0.3721
	MnO_3	O	0.4706
	Mn_2O_7	O	0.5091
	HgO	O	0.0741
	Hg_2O	O	0.0385
	KO	O	0.1706
	SiO_2	O	0.5333
	AgO	O	0.0690
	NaO	O	0.2581
	SrO	O	0.1546
	SnO	O	0.2162
	ZnO	O	0.1976
Potassium	KO	K	0.8294
	KO, SO_3	KO	0.5402
	KO, NO_5	KO	0.4654
	KCl	KO	0.6309
	KCl	K	0.6235
	$KCl, PtCl_2$	KO	0.1922
	$KCl, PtCl_2$	KCl	0.3047
Rubidium	$RbCl$	RbO	0.7725
	RbO	Rb	0.9143
Silicon	SiO_2	Si	0.4667
Silver	$AgCl$	Ag	0.7526
	$AgCl$	AgO	0.8084
Sodium	NaO	Na	0.7419
	NaO, SO_3	NaO	0.4366
	NaO, NO_5	NaO	0.3647
	$NaCl$	NaO	0.5299
	$NaCl$	Na	0.3922
	NaO, CO_2	NaO	0.5849
Strontium	SrO	Sr	0.8454
	SrO, SO_3	SrO	0.5640
	SrO, CO_2	SrO	0.7015

Element.	Found.	Sought.	Factor.
Sulphur	BaO, SO_3	S	0.1372
	BaO, SO_3	SO_3	0.3433
	AsS_3	S	0.3902
Tantalum	TaO_5	Ta	0.8113
Thallium	TlS	Tl	0.9272
"	TlI	Tl	0.6163
Thorium	ThO_2	Th	0.8788
Tin	SnO_2	Sn	0.7838
	SnO_2	SnO	0.8919
Titanium	TiO_2	Ti	0.6098
Tungsten	WO_3	W	0.7143
Yttrium	YO	Y	0.7941
Zinc	ZnO	Zn	0.8024
Zirconium	ZrO_2	Zr	0.8485

FORMULAS FOR INDIRECT ESTIMATION.

1. *Estimation of Sulphates of Potassium and Sodium.*

$$N = 9.65158 \, A - 4.43751 \, S$$

$$K = S - N$$

S = weight of both Sulphates
K = weight of KO, SO_3
N = weight of NaO, SO_3
A = weight of SO_4.

2. *Estimation of Potassium and Sodium from the Chlorides.*

$$K = 2.4375 \, S - 4.01672 \, A$$
$$N = S - (A + K)$$

S = weight of both Chlorides
A = weight of Chlorine
N = weight of Sodium
K = weight of Potassium.

3. *Carbonates of Lime and Strontia.*

$$Ca = 7.0574 \, A - 2.10526 \, S$$
$$Sr = S - Ca$$

S = weight of both Carbonates
A = weight of Carbonic Acid
Ca = weight of Carbonate of Lime
Sr = weight of Carbonate of Strontia.

TABLE OF FACTORS.

FOR SODA, POTASSA, AND BINOXIDE OF MANGANESE.

The weight of the substance found multiplied by the factor gives the weight of the substance sought.

Found.	Sought.	Factor.
Soda.		
CO_2	NaO	1.4091
"	NaO, HO	1.8182
"	NaO, CO_2	2.4091
"	$NaO, CO_2 + 10HO$	6.5000
NaO	NaO, HO	1.2903
"	NaO, CO_2	1.7097
"	$NaO, CO_2 + 10HO$	4.6129
Potassa.		
CO_2	KO	2.1363
"	KO, HO	2.5450
"	KO, CO_2	3.1363
KO	KO, HO	1.1915
"	KO, CO_2	1.4681
Manganese.		
CO_2	MnO_2	0.9773

TABLE

Giving the corresponding per cents of Chlorine and Binoxide of Manganese, from the amount of Sulphate of Iron Peroxidized, when 5 grammes of Black Oxide of Manganese and 31.9 grammes of Sulphate of Iron are taken for testing.

OTTO'S LEHRBUCH.

Decigrammes remaining of Sulphate of Iron.	Decigrammes used of Sulphate of Iron.	Per cent of Binoxide contained in the Manganese.	Per cent of Chlorine liberated by the Manganese.	Decigrammes remaining of Sulphate of Iron.	Decigrammes used of Sulphate of Iron.	Per cent of Binoxide contained in the Manganese.	Per cent of Chlorine liberated by the Manganese.
0	319	100	81.3	98	221	69	56.3
3	316	99	80.6	102	217	68	55.3
6	313	98	79.8	105	214	67	54.5
9	310	97	79.0	108	211	66	53.8
12	307	96	78.2	111	208	65	53.0
15	304	95	77.4	115	204	64	52.0
18	301	94	76.5	118	201	63	51.2
22	297	63	75.7	121	198	62	50.5
25	294	92	75.0	124	195	61	49.7
28	291	91	74.3	127	192	60	48.9
31	288	90	73.5	130	189	59	48.2
34	285	89	72.7	134	185	58	47.2
38	281	88	71.6	137	182	57	46.4
41	278	87	70.9	140	179	56	45.6
44	275	86	70.1	143	176	55	44.9
47	272	85	69.3	147	172	54	43.8
50	269	84	68.6	150	169	53	43.1
53	266	83	67.8	153	166	52	42.3
57	262	82	66.8	156	163	51	41.5
60	259	81	66.0	159	160	50	40.8
63	256	80	65.3	162	157	49	40.0
66	253	79	64.5	165	154	48	39.3
69	250	78	63.7	168	151	47	38.5
73	246	77	62.7	172	147	46	37.5
76	243	76	61.9	175	144	45	36.7
79	240	75	61.2	178	141	44	35.9
82	237	74	60.4	182	137	43	34.9
85	234	73	59.6	185	134	42	34.2
89	230	72	58.6	188	131	41	33.4
92	227	71	57.8	191	128	40	32.6
95	224	70	57.1				

TABLE

Giving the per cent of Chlorine in Chloride of Lime from the number of parts it takes to Peroxidize 3.9 grammes of Sulphate of Iron, when 100 parts of the Solution contain 5 grammes of the Chloride of Lime to be tested.

(OTTO'S LEHRBUCH.)

Number of Parts used of the Solution of Chloride of Lime.	Per cent of Chlorine.	Number of Parts used of the Solution of Chloride of Lime.	Per cent of Chlorine.	Number of Parts used of the Solution of Chloride of Lime.	Per cent of Chlorine.
31	32.2	55	18.2	79	12.7
32	31.2	56	17.8	80	12.5
33	30.3	57	17.5	81	12.3
34	29.4	58	17.2	82	12.2
35	28.6	59	17.0	83	12.0
36	27.8	60	16.7	84	11.9
37	27.0	61	16.4	85	11.7
38	26.3	62	16.1	86	11.6
39	25.6	63	15.9	87	11.5
40	25.0	64	15.6	88	11.3
41	24.4	65	15.4	89	11.2
42	23.8	66	15.1	90	11.1
43	23.3	67	14.9	91	10.9
44	22.7	68	14.7	92	10.8
45	22.2	69	14.5	93	10.7
46	21.7	70	14.3	94	10.6
47	21.3	71	14.1	95	10.5
48	20.8	72	13.9	96	10.4
49	20.4	73	13.7	97	10.3
50	20.0	74	13.5	98	10.2
51	19.6	75	13.3	99	10.1
52	19.2	76	13.1	100	10.0
53	18.8	77	13.0		
54	18.5	78	12.8		

TABLES

RELATING TO

SPECIFIC GRAVITY.

SPECIFIC GRAVITIES OF THE ELEMENTS.

Element.	Specific Gravity.	Element.	Specific Gravity.
		Palladium	11–12
Solids and Liquids.		Phosphorus { Amorphous	2.089
Water = 1 at 0° C., 760 m. Barom.		Phosphorus { Crystallized	1.826
		Platinum	20.8–21.74
Aluminum	2.670	Potassium	0.865
Antimony	6.720	Rhodium	11.000
Arsenic	5.630	Rubidium	1.520
Barium	1.850	Ruthenium	11–12
Bismuth	9.780	Selenium { Amorphous	4.250
Boron	2.680	Selenium { Crystallized	4.800
Bromine	3.187	Silicon	2.490
Cadmium	8.604	Silver	10.570
Calcium	1.578	Sodium	0.972
Carbon { Diamond	3.500	Strontium	2.540
Carbon { Graphite	2.200	Sulphur { Rhombic	2.070
Carbon { Coal of Sugar	2.000	Sulphur { Monoclinic	1.958
Chromium	6.810	Sulphur { Amorphous	1.919
Cobalt	8.950	Tantalum	10.780
Copper	8.8–9	Tellurium	6.650
Glucinum	2.100	Thallium	11.862
Gold	19.340	Tin	7.300
Indium	7.200	Tungsten	17.500
Iodine	4.950	Uranium	18.400
Iridium	21.800	Vanadium	3.640
Iron	7.840	Zinc	7.200
Lead	11.330		
Lithium	0.593	**Gases.**	
Magnesium	1.750	Air = 1 at 0° C., 760 m. Barom.	
Manganese	8.000		
Mercury	13.596	Chlorine	2.47000
Molybdenum	8.640	Hydrogen	0.06926
Nickel	8.900	Nitrogen	0.97137
Osmium	21.400	Oxygen	1.10563

TABLE

OF THE SPECIFIC GRAVITIES OF THE MOST FREQUENTLY OCCURRING
COMBINATIONS.

BOEDEKER. — Supplement zu den Lehrbüchern der Chemie und Mineralogie.

Name.			Formula.	Specific Gravity.
Bromides.				
Ammonium, Bromide			NH_4Br	2.266
Barium	"		$BaBr$	4.230
"	"	Crystal.	$BaBr + 2HO$	3.690
Mercury	"		$HgBr$	5.920
"	"	Sub	Hg_2Br	7.307
Potassium	"		KBr	2.415
Silver	"		$AgBr$	6.273
Sodium	"		$NaBr$	3.079
Chlorides.				
Ammonium, Chloride			NH_4Cl	1.500
Barium	"		$BaCl$	3.800
"	"	Crystal.	$BaCl + 2HO$	3.050
Calcium	"		$CaCl$	2.205
"	"	Crystal.	$CaCl + 6HO$	1.612
Copper	"		Cu_2Cl	3.680
Iron	"		$FeCl$	2.528
Lead	"		$PbCl$	5.540
Lithium	"		LCl	1.998
Manganese	"		$MnCl + 4HO$	2.0146
Mercury	"		$HgCl$	5.420
"	"	Sub	Hg_2Cl	6.990
Nickel	"		$NiCl$	2.560
Potassium	"		KCl	1.950
Silver	"		$AgCl$	5.500
Sodium	"		$NaCl$	2.160
Strontium	"		$SrCl$	2.960
Tin	"		$SnCl + 2HO$	2.293
Iodides.				
Barium, Iodide			BaI	4.917
Copper	"		CuI	4.410
Lead	"		PbI	6.384

Name.	Formula.	Specific Gravity.
Mercury, Iodide	HgI	6.257
" " Sub	Hg_2I	7.750
Potassium "	KI	3.056
Silver "	AgI	5.614
Sodium "	NaI	3.450

Oxides.

Name.	Formula.	Specific Gravity.
Aluminum, Oxide	Al_2O_3	4.000
Antimonic Acid	SbO_5	6.525
Antimonous Acid	SbO_4	6.695
Antimony, Oxide	SbO_3	5.566
Arsenous Acid	AsO_3	3.884
Arsenic "	AsO_5	3.734
Barium, Oxide	BaO	5.456–4.732
" " Crystal.	$BaO + 9HO$	1.656
" Hydrate	BaO, HO	4.495
Boric Acid	BO_3	1.830
" " Crystal.	$BO_3 + 3HO$	1.479
Bismuth, Oxide	BiO_3	8.968
Cadmium "	CdO	8.110
Calcium "	CaO	3.180
Chromium "	Cr_2O_3	5.210
Cobalt "	Co_2O_3	5.600
Copper "	CuO	6.430
" " Sub	Cu_2O	5.751
Iron, Protosesquioxide	Fe_3O_4	5.185
" Sesquioxide	Fe_2O_3	5.224
" Hydrate	Fe_2O_3, HO	3.7–3.9
Lead, Oxide	PbO	9.360
" " Per	PbO_2	9.300
Magnesium, Oxide	MgO	3.200
Manganese "	MnO	4.720
" " Sesqui	Mn_2O_3	4.820
" " Per	MnO_2	4.900
" " Hydrate	Mn_2O_3, HO	4.328
Mercury, Oxide	HgO	11.290
" " Sub	Hg_2O	8.950
Molybdenum, Oxide	MoO_2	5.606
" Acid	MoO_3	3.460
Nickel, Oxide	NiO	6.660
" " Sesqui	Ni_2O_3	4.840
Potassium, Oxide	KO	2.656
" Hydrate	KO, HO	2.100
Silicic Acid	SiO_2	2.660
Sodium, Oxide	NaO	2.805
" Hydrate	NaO, HO	2.000

Name.	Formula.	Specific Gravity.
Strontium, Oxide	SrO	4.611
" " Crystal	$SrO + 9HO$	1.396
Sulphuric Acid	SO_3	1.970 (20°)
Tin, Oxide	SnO	6.666
" " Bin	SnO_2	6.720–6.950
Titanic Acid ⎰ Anatase	TiO_2	3.840
Titanic Acid ⎰ Brookite	TiO_2	4.130
⎱ Rutile	TiO_2	4.250
Tungstic Acid	WO_3	7.1369
Uranium, Oxide	UO	10.150
Uranium Protosesquioxide	U_2O_4	7.310
Vanadic Acid	VO_3	3.510
Zinc, Oxide	ZnO	5.630

Sulphides.

Name.	Formula.	Specific Gravity.
Antimony, Tersulphide	SbS_3	4.752
Arsenic "	AsS_3	3.480
" Bisulphide	AsS_2	3.544
Bismuth, Sulphide	BiS_2	7.000
Cadmium "	CdS	4.900
Cobalt "	Co_2S_3	4.800
Copper "	CuS	4.163
" " Sub	Cu_2S	5.500
Iron " Bi	FeS_2	4.65–5.1
" "	FeS	4.400
Lead "	PbS	6.924
Manganese "	MnS	4.000
" " Bi	MnS_2	3.463
Mercury "	HgS	8.124
Molybdenum "	MoS_2	4.690
Nickel "	NiS	5.650
Potassium "	KS	2.130
Silver "	AgS	6.850
Sodium "	NaS	2.471
Tin "	SnS	4.973
" " Bi	SnS_2	4.600
Zinc "	ZnS	3.920

Borates.

Name.	Formula.	Specific Gravity.
Borax	$NaO, 2BO_3 + 10HO$	1.740
" Anhydrous	$NaO, 2BO_3$	2.600

Name.	Formula.	Specific Gravity.
Carbonates.		
Ammonium, Carbonate	$2NH_4O, 3CO_2$	1.450
Barium "	BaO, CO_2	4.300
Calcium " Arragonite	CaO, CO_2	2.900
" " Calc-spar	CaO, CO_2	2.720
Iron "	FeO, CO_2	3.870
Lead "	PbO, CO_2	6.400
Lithium "	LO, CO_2	2.110
Magnesium "	MgO, CO_2	2.990
Manganese "	MnO, CO_2	3.55–3.95
Potassium "	KO, CO_2	2.267
Sodium "	NaO, CO_2	2.509
" " Crystal.	$NaO, CO_2 + 10HO$	1.454
Strontium "	SrO, CO_2	3.600
Zinc "	ZnO, CO_2	4.1–4.5
Chlorates.		
Barium, Chlorate	BaO, ClO_5	2.988
Potassium "	KO, ClO_5	2.350
Sodium "	NaO, ClO_5	2.289
Silver "	AgO, ClO_5	4.430
Chromates.		
Ammonium, Chromate, Bi	$NH_4O, 2CrO_3$	2.367
Lead "	PbO, CrO_3	6.100
Potassium "	KO, CrO_3	2.640
" " Bi	$KO, 2CrO_3$	2.603
Nitrates.		
Ammonium, Nitrate	NH_4O, NO_5	1.740
Barium "	BaO, NO_5	3.200
Calcium "	CaO, NO_5	2.240
Lead "	PbO, NO_5	4.470
Lithium "	LO, NO_5	2.334
Potassium "	KO, NO_5	2.058
Silver "	AgO, NO_5	4.355
Strontium "	SrO, NO_5	2.810
Sodium "	NaO, NO_5	2.200
Oxalates.		
Ammonium, Oxalate	$NH_4O, \overline{O} + 2HO$	1.500
Potassium "	$KO, \overline{O} + HO$	2.120
" " Acid	$KO, HO, 2\overline{O} + 2HO$	2.044

Name.	Formula.	Specific Gravity.
Phosphates.		
Lead, Phosphate	$3PbO, PO_5$	7.208
Sodium "	$2NaO, HO, PO_5 + 24HO$	1.525
" " Pyro	$2NaO, PO_5 + 10HO$	1.836
Silver "	$3AgO, PO_5$	5.306
" " Pyro	$2AgO, PO_5$	7.300
Sulphates.		
Ammonium, Sulphate	NH_4O, SO_3	1.760
Barium "	BaO, SO_3	4.500
Calcium "	CaO, SO_3	2.960
" " Crystal.	$CaO, SO_3 + 2HO$	2.330
Cobalt " "	$CoO, SO_3 + 7HO$	1.924
Copper "	CuO, SO_3	3.572
" " Crystal.	$CuO, SO_3 + 5HO$	2.300
Iron "	FeO, SO_3	2.840
" " Crystal.	$FeO, SO_3 + 7HO$	1.8–1.9
Lead "	PbO, SO_3	6.300
Lithium, Sulphate	LO, SO_3	2.210
" " Crystal.	$LO, SO_3 + HO$	2.020
Magnesia "	MgO, SO_3	2.600
" " Crystal.	$MgO, SO_3 + 7HO$	1.750
Manganese "	MnO, SO_3	3.100
" " Crystal.	$MnO, SO_3 + 4HO$	2.090
Nickel " "	$NiO, SO_3 + 7HO$	2.040
Potassium "	KO, SO_3	2.660
" " Acid	$KO, 2SO_3$	2.277
Potassa, Alum	$KO, SO_3, Al_2O_3, 3SO_3 + 24HO$	1.724
Silver, Sulphate	AgO, SO_3	5.400
Sodium "	NaO, SO_3	2.460
" " Crystal.	$NaO, SO_3 + 10HO$	1.500
Soda, Alum	$NaO, SO_3, Al_2O_3, 3SO_3 + 24HO$	1.880
Strontium, Sulphate	SrO, SO_3	3.925
Zinc "	ZnO, SO_3	3.400
" Crystal.	$ZnO, SO_3 + 7HO$	1.9–2.1

TABLE

OF THE PER CENT OF DRY AMMONIA GAS IN A SOLUTION AT 14° C.

CARIUS. — Ann. Ch. u. Pharm. 1856, 99, pp. 163, 164.

Specific Gravity.	Per cent Ammonia.	Specific Gravity.	Per cent Ammonia.	Specific Gravity.	Per cent Ammonia.	Specific Gravity.	Per cent Ammonia.
0.8844	36	0.9052	27	0.9314	18	0.9631	9
0.8864	35	0.9078	26	0.9347	17	0.9670	8
0.8885	34	0.9106	25	0.9380	16	0.9709	7
0.8907	33	0.9133	24	0.9414	15	0.9749	6
0.8929	32	0.9162	23	0.9449	14	0.9790	5
0.8953	31	0.9191	22	0.9484	13	0.9831	4
0.8976	30	0.9221	21	0.9520	12	0.9873	3
0.9001	29	0.9251	20	0.9556	11	0.9915	2
0.9026	28	0.9283	19	0.9593	10	0.9959	1

TABLE

OF THE PER CENT OF DRY AMMONIA GAS IN A SOLUTION AT 16° C.

OTTO'S LEHRBUCH.

Specific Gravity.	Per cent Ammonia.	Specific Gravity.	Per cent Ammonia.	Specific Gravity.	Per cent Ammonia.	Specific Gravity.	Per cent Ammonia.
0.9517	12.000	0.9583	10.250	0.9654	8.375	0.9726	6.500
0.9521	11.875	0.9588	10.125	0.9659	8.250	0.9730	6.375
0.9526	11.750	0.9593	10.000	0.9664	8.125	0.9735	6.250
0.9531	11.625	0.9597	9.875	0.9669	8.000	0.9740	6.125
0.9536	11.500	0.9602	9.750	0.9673	7.875	0.9745	6.000
0.9540	11.375	0.9607	9.625	0.9678	7.750	0.9749	5.875
0.9545	11.250	0.9612	9.500	0.9683	7.625	0.9754	5.750
0.9550	11.125	0.9616	9.375	0.9688	7.500	0.9759	5.625
0.9555	11.000	0.9621	9.250	0.9692	7.375	0.9764	5.500
0.9556	10.950	0.9626	9.125	0.9697	7.250	0.9768	5.375
0.9559	10.875	0.9631	9.000	0.9702	7.125	0.9773	5.250
0.9564	10.750	0.9636	8.875	0.9707	7.000	0.9778	5.125
0.9569	10.625	0.9641	8.750	0.9711	6.875	0.9783	5.000
0.9574	10.500	0.9645	8.625	0.9716	6.750		
0.9578	10.375	0.9650	8.500	0.9721	6.625		

TABLE

OF THE PER CENT OF POTASSA (KO) IN A SOLUTION AT 15° C.

TÜNNERMANN. — Handwörterbuch, Bd. IV. pp. 253.

Specific Gravity.	Per cent Potassa.	Specific Gravity.	Per cent Potassa.	Specific Gravity.	Per cent Potassa.
1.3300	28.290	1.1979	18.671	1.0819	8.487
1.3131	27.158	1.1839	17.540	1.0703	7.355
1.2966	26.027	1.1702	16.408	1.0589	6.224
1.2803	24.895	1.1568	15.277	1.0478	5.002
1.2648	23.764	1.1437	14.145	1.0369	3.961
1.2493	22.632	1.1308	13.013	1.0260	2.829 .
1.2342	21.500	1.1182	11.882	1.0153	1.697
1.2268	20.935	1.1059	10.750	1.0050	0.5658
1.2122	19.803	1.0938	9.619		

TABLE

OF THE PER CENT OF SODA (NaO) IN A SOLUTION AT 15° C.

TÜNNERMANN. — Handwörterbuch, Bd. V. pp. 522.

Specific Gravity.	Per cent Soda.	Specific Gravity.	Per cent Soda.	Specific Gravity.	Per cent Soda.
1.4285	30.220	1.2982	20.550	1.1528	10.275
1.4193	29.616	1.2912	19.945	1.1428	9.670
1.4101	29.011	1.2843	19.341	1.1330	9.066
1.4011	28.407	1.2775	18.730	1.1233	8.462
1.3923	27.802	1.2708	18.132	1.1137	7.857
1.3836	27.200	1.2642	17.528	1.1042	7.253
1.3751	26.594	1.2578	16.923	1.0948	6.648
1.3668	25.989	1.2515	16.319	1.0855	6.044
1.3586	25.385	1.2453	15.714	1.0764	5.440
1.3505	24.780	1.2392	15.110	1.0675	4.835
1.3426	24.176	1.2280	14.506	1.0587	4.231
1.3349	23.572	1.2178	13.901	1.0500	3.626
1.3273	22.967	1.2058	13.297	1.0414	3.022
1.3198	22.363	1.1948	12.692	1.0330	2.418
1.3143	21.894	1.1841	12.088	1.0246	1.813
1.3125	21.758	1.1734	11.484	1.0163	1.209
1.3053	21.154	1.1630	10.879	1.0081	0.604

TABLE

OF THE SPECIFIC GRAVITY OF SULPHURIC ACID.

Calculated by Otto from Bineau's data for the Temperature of 15° *C.*

Specific Gravity.	Hy-drated Acid.	SO_3	Specific Gravity.	Hy-drated Acid.	SO_3	Specific Gravity.	Hy-drated Acid.	SO_3
1.8426	100	81.63	1.5780	66	53.87	1.2390	32	26.12
1.8420	99	80.81	1.5570	65	53.05	1.2310	31	25.30
1.8406	98	80.00	1.5450	64	52.24	1.2230	30	24.49
1.8400	97	79.18	1.5340	63	51.42	1.2150	29	23.67
1.8384	96	78.36	1.5230	62	50.61	1.2066	28	22.85
1.8376	95	77.55	1.5120	61	49.79	1.1980	27	22.03
1.8356	94	76.73	1.5010	60	48.98	1.1900	26	21.22
1.8340	93	75.91	1.4900	59	48.16	1.1820	25	20.40
1.8310	92	75.10	1.4800	58	47.34	1.1740	24	19.58
1.8270	91	74.28	1.4690	57	46.53	1.1670	23	18.77
1.8220	90	73.47	1.4586	56	45.71	1.1590	22	17.95
1.8160	89	72.65	1.4480	55	44.89	1.1516	21	17.14
1.8090	88	71.83	1.4380	54	44.07	1.1440	20	16.32
1.8020	87	71.02	1.4280	53	43.26	1.1360	19	15.51
1.7940	86	70.10	1.4180	52	42.45	1.1290	18	14.69
1.7860	85	69.38	1.4080	51	41.63	1.1210	17	13.87
1.7770	84	68.57	1.3980	50	40.81	1.1136	16	13.06
1.7670	83	67.75	1.3886	49	40.00	1.1060	15	12.24
1.7560	82	66.94	1.3790	48	39.18	1.0980	14	11.42
1.7450	81	66.12	1.3700	47	38.36	1.0910	13	10.61
1.7340	80	65.30	1.3610	46	37.55	1.0830	12	9.79
1.7220	79	64.48	1.3510	45	36.73	1.0756	11	8.98
1.7100	78	63.67	1.3420	44	35.82	1.0680	10	8.16
1.6980	77	62.85	1.3330	43	35.10	1.0610	9	7.34
1.6860	76	62.04	1.3240	42	34.28	1.0536	8	6.53
1.6750	75	61.22	1.3150	41	33.47	1.0464	7	5.71
1.6630	74	60.40	1.3060	40	32.65	1.0390	6	4.89
1.6510	73	59.59	1.2976	39	31.83	1.0320	5	4.08
1.6390	72	58.77	1.2890	38	31.02	1.0256	4	3.26
1.6370	71	57.95	1.2810	37	30.20	1.0190	3	2.445
1.6150	70	57.14	1.2720	36	29.38	1.0130	2	1.63
1.6040	69	56.32	1.2640	35	28.57	1.0064	1	0.816
1.5920	68	55.59	1.2560	34	27.75			
1.5300	67	54.69	1.2476	33	26.94			

TABLE

OF THE QUANTITY OF SULPHURIC ACID OF 1.86 SPECIFIC GRAVITY
TO BE ADDED TO 100 PARTS OF WATER TO PRODUCE AN ACID OF
A GIVEN SPECIFIC GRAVITY.

ANTHON.

Parts Acid.	Specific Gravity.	Parts Acid.	Specific Gravity.	Parts Acid.	Specific Gravity.
1	1.009	130	1.456	370	1.723
2	1.015	140	1.473	380	1.727
5	1.035	150	1.490	390	1.730
10	1.060	160	1.510	400	1.733
15	1.090	170	1.530	410	1.737
20	1.113	180	1.543	420	1.740
25	1.140	190	1.556	430	1.743
30	1.165	200	1.568	440	1.746
35	1.187	210	1.580	450	1.750
40	1.210	220	1.593	460	1.754
45	1.229	230	1.606	470	1.757
50	1.248	240	1.620	480	1.760
55	1.265	250	1.630	490	1.763
60	1.280	260	1.640	500	1.766
65	1.297	270	1.648	510	1.768
70	1.312	280	1.654	520	1.770
75	1.326	290	1.667	530	1.772
80	1.340	300	1.678	540	1.774
85	1.357	310	1.689	·550	1.776
90	1.372	320	1.700	560	1.777
95	1.386	330	1.705	580	1.778
100	1.398	340	1.710	590	1.780
110	1.420	350	1.714	600	1.782
120	1.438	360	1.719		

TABLE

OF THE SPECIFIC GRAVITY OF NITRIC ACID AT 15° C.

URE. — Dict. Arts.

Specific Gravity.	Per cent of NO_5	Specific Gravity.	Per cent of NO_5	Specific Gravity.	Per cent of NO_5
1.500	79.7	1.378	52.6	1.190	26.3
1.498	78.9	1.373	51.8	1.183	25.5
1.496	78.1	1.368	51.1	1.177	24.7
1.494	77.3	1.363	50.2	1.171	23.9
1.491	76.5	1.358	49.4	1.165	23.1
1.488	75.7	1.353	48.6	1.159	22.3
1.485	74.9	1.348	47.8	1.153	21.5
1.482	74.1	1.343	47.0	1.147	20.7
1.479	73.3	1.338	46.2	1.140	19.9
1.476	72.5	1.332	45.4	1.134	19.1
1.473	71.7	1.327	44.6	1.129	18.3
1.470	70.9	1.322	43.8	1.123	17.5
1.467	70.1	1.316	43.0	1.117	16.7
1.464	69.3	1.311	42.2	1.111	15.9
1.460	68.5	1.306	41.4	1.105	15.1
1.457	67.7	1.300	40.6	1.099	14.3
1.453	66.9	1.295	39.9	1.094	13.5
1.450	66.2	1.289	39.1	1.088	12.7
1.446	65.3	1.283	38.3	1.082	12.0
1.442	64.6	1.277	37.5	1.076	11.2
1.439	63.8	1.271	36.7	1.071	10.4
1.435	63.0	1.264	35.9	1.065	9.6
1.431	62.2	1.258	35.1	1.059	8.8
1.427	61.4	1.252	34.3	1.054	8.0
1.423	60.6	1.246	33.5	1.048	7.2
1.419	59.8	1.240	32.7	1.043	6.4
1.415	59.0	1.234	31.9	1.038	5.6
1.411	58.2	1.228	31.1	1.032	4.8
1.407	57.4	1.221	30.3	1.027	4.0
1.402	56.6	1.215	29.5	1.021	3.2
1.398	55.8	1.208	28.7	1.016	2.4
1.395	55.0	1.202	27.9	1.011	1.6
1.388	54.2	1.196	27.1	1.005	0.8
1.383	53.4				

TABLE

OF THE SPECIFIC GRAVITY OF HYDROCHLORIC ACID AT 15° C.

URE. — Dict. Arts.

Specific Gravity.	Per cent Hydrochloric Acid Gas.	Specific Gravity.	Per cent Hydrochloric Acid Gas.	Specific Gravity.	Per cent Hydrochloric Acid Gas.
1.2000	40.777	1.1328	26.913	1.0657	13.456
1.1982	40.369	1.1308	26.505	1.0637	13.049
1.1964	39.961	1.1287	26.098	1.0617	12.641
1.1946	39.554	1.1267	25.690	1.0597	12.233
1.1928	39.146	1.1247	25.282	1.0577	11.825
1.1910	38.738	1.1226	24.874	1.0557	11.418
1.1893	38.330	1.1206	24.466	1.0537	11.010
1.1875	37.923	1.1185	24.058	1.0517	10.602
1.1857	37.516	1.1164	23.650	1.0497	10.194
1.1846	37.108	1.1143	23.242	1.0477	9.786
1.1822	36.700	1.1123	22.834	1.0457	9.379
1.1802	36.292	1.1102	22.426	1.0437	8.971
1.1782	35.884	1.1082	22.019	1.0417	8.563
1.1762	35.476	1.1061	21.611	1.0397	8.155
1.1741	35.068	1.1041	21.203	1.0377	7.747
1.1721	34.660	1.1020	20.796	1.0357	7.340
1.1701	34.252	1.1000	20.388	1.0337	6.932
1.1681	33 845	1.0980	19.980	1.0318	6.524
1.1661	33.437	1.0960	19.572	1.0298	6.116
1.1641	33.029	1.0939	19.165	1.0279	5.709
1.1620	32.621	1.0919	18.757	1.0259	5.301
1.1599	32 213	1.0899	18.349	1.0239	4.893
1.1578	31.805	1.0879	17.941	1.0220	4.486
1.1557	31.398	1.0859	17.534	1.0200	4.078
1.1537	30.990	1.0838	17.126	1.0180	3.670
1.1515	30.582	1.0818	16.718	1.0160	3.262
1.1494	30.174	1.0798	16.310	1.0140	2.854
1.1473	29.767	1.0778	15.902	1.0120	2.447
1.1452	29.359	1.0758	15.494	1.0100	2.039
1.1431	28.951	1.0738	15.087	1.0080	1.631
1.1410	28.544	1.0718	14.679	1.0060	1.124
1.1389	28.136	1.0697	14.271	1.0040	0.816
1.1369	27.728	1.0677	13.863	1.0020	0.408
1.1349	27.321				

TABLE

OF THE SPECIFIC GRAVITY OF HYDRATED ACETIC ACID.

MOHR. — Ann. Pharm. 1839, **31**, s. 284.

Specific Gravity.	Per cent Hydrated Acetic Acid.	Specific Gravity.	Per cent Hydrated Acetic Acid.	Specific Gravity.	Per cent Hydrated Acetic Acid.	Specific Gravity.	Per cent Hydrated Acetic Acid.
1.0635	100	1.0720	75	1.0600	50	1.0340	25
1.0655	99	1.0720	74	1.0590	49	1.0330	24
1.0670	98	1.0720	73	1.0580	48	1.0320	23
1.0680	97	1.0710	72	1.0560	47	1.0310	22
1.0690	96	1.0710	71	1.0550	46	1.0290	21
1.0700	95	1.0700	70	1.0550	45	1.0270	20
1.0706	94	1.0700	69	1.0540	44	1.0260	19
1.0708	93	1.0700	68	1.0530	43	1.0250	18
1.0716	92	1.0690	67	1.0520	42	1.0240	17
1.0721	91	1.0690	66	1.0515	41	1.0230	16
1.0730	90	1.0680	65	1.0513	40	1.0220	15
1.0730	89	1.0680	64	1.0500	39	1.0200	14
1.0730	88	1.0680	63	1.0490	38	1.0180	13
1.0730	87	1.0670	62	1.0480	37	1.0170	12
1.0730	86	1.0670	61	1.0470	36	1.0160	11
1.0730	85	1.0670	60	1.0460	35	1.0150	10
1.0730	84	1.0660	59	1.0450	34	1.0130	9
1.0730	83	1.0660	58	1.0440	33	1.0120	8
1.0730	82	1.0650	57	1.0424	32	1.0100	7
1.0732	81	1.0640	56	1.0410	31	1.0080	6
1.0735	80	1.0640	55	1.0400	30	1.0067	5
1.0735	79	1.0630	54	1.0390	29	1.0055	4
1.0732	78	1.0630	53	1.0380	28	1.0040	3
1.0732	77	1.0620	52	1.0360	27	1.0020	2
1.0730	76	1.0610	51	1.0350	26	1.0010	1

When water is added to Hydrated Acetic Acid the Specific Gravity of the fluid does not fall; on the contrary, it rises until 3 atoms or 51.5 per cent have been added.

In reference to this, Thompson gives the following Table.

EQUIVALENTS OF ACID AND WATER AT 15° C.

Acid.	Water.	Specific Gravity.	Acid.	Water.	Specific Gravity.
1	1	1.06296	1	6	1.06708
1	2	1.07060	1	7	1.06349
1	3	1.07084	1	8	1.05974
1	4	1.07132	1	9	1.05794
1	5	1.06820	1	10	1.05439

TABLE

OF THE PROPORTION BY WEIGHT OF ABSOLUTE ALCOHOL IN SPIRITS
OF DIFFERENT SPECIFIC GRAVITIES AT 15°.5 C.

FOWNES'S Chem., p. 579.

Specific Gravity.	Per cent.	Specific Gravity.	Per cent.	Specific Gravity.	Per cent.
0.9991	½	0.9511	34	0.8769	68
0.9981	1	0.9490	35	0.8745	69
0.9965	2	0.9470	36	0.8721	70
0.9947	3	0.9452	37	0.8696	71
0.9930	4	0.9434	38	0.8672	72
0.9914	5	0.9416	39	0.8649	73
0.9898	6	0.9396	40	0.8625	74
0.9884	7	0.9376	41	0.8603	75
0.9869	8	0.9356	42	0.8581	76
0.9855	9	0.9335	43	0.8557	77
0.9841	10	0.9314	44	0.8533	78
0.9828	11	0.9292	45	0.8508	79
0.9815	12	0.9270	46	0.8483	80
0.9802	13	0.9249	47	0.8459	81
0.9789	14	0.9228	48	0.8434	82
0.9778	15	0.9206	49	0.8408	83
0.9766	16	0.9184	50	0.8382	84
0.9753	17	0.9160	51	0.8357	85
0.9741	18	0.9135	52	0.8331	86
0.9728	19	0.9113	53	0.8305	87
0.9716	20	0.9090	54	0.8279	88
0.9704	21	0.9069	55	0.8254	89
0.9691	22	0.9047	56	0.8228	90
0.9678	23	0.9025	57	0.8199	91
0.9665	24	0.9001	58	0.8172	92
0.9652	25	0.8979	59	0.8145	93
0.9638	26	0.8956	60	0.8118	94
0.9623	27	0.8932	61	0.8089	95
0.9609	28	0.8908	62	0.8061	96
0.9593	29	0.8886	63	0.8031	97
0.9578	30	0.8863	64	0.8001	98
0.9560	31	0.8840	65	0.7969	99
0.9544	32	0.8816	66	0.7938	100
0.9528	33	0.8793	67		

TABLE

OF THE SPECIFIC GRAVITY OF A MIXTURE OF ALCOHOL AND WATER
BY VOLUME.

Temperature = 15°.56 *C.*   *Water* = 1 at 3°.9 *C.*

TRALLES. — Gilbert's Annalen der Phys., 1811, 38, s. 368.

Per cent Alcohol.	Specific Gravity.	Per cent Alcohol.	Specific Gravity.	Per cent Alcohol.	Specific Gravity.
1	9976	35	9583	68	8941
2	9961	36	9570	69	8917
3	9947	37	9559	70	8892
4	9933	38	9541	71	8867
5	9919	39	9526	72	8842
6	9906	40	9510	73	8817
7	9893	41	9494	74	8791
8	9881	42	9478	75	8765
9	9869	43	9461	76	8739
10	9857	44	9444	77	8712
11	9845	45	9427	78	8685
12	9834	46	9409	79	8658
13	9823	47	9391	80	8631
14	9812	48	9373	81	8603
15	9802	49	9354	82	8575
16	9791	50	9335	83	8547
17	9781	51	9315	84	8518
18	9771	52	9295	85	8488
19	9761	53	9275	86	8458
20	9751	54	9254	87	8428
21	9741	55	9234	88	8397
22	9731	56	9213	89	8365
23	9720	57	9192	90	8332
24	9710	58	9170	91	8299
25	9700	59	9148	92	8265
26	9689	60	9126	93	8230
27	9679	61	9104	94	8194
28	9668	62	9082	95	8157
29	9657	63	9059	96	8118
30	9646	64	9036	97	8077
31	9634	65	9013	98	8034
32	9622	66	8989	99	7988
33	9609	67	8965	100	7939
34	9596				

SPECIFIC GRAVITIES.

TABLE

FOR THE MIXING OF WATER AND ALCOHOL ACCORDING TO VOLUME, AT 17°.5 C.

MEISSNER'S Aräometrie.

Volumes Alcohol.	Volumes Water.	Specific Gravity.	The Volume becomes after mixing.	Volumes Alcohol.	Volumes Water.	Specific Gravity.	The Volume becomes after mixing.	Volumes Alcohol.	Volumes Water.	Specific Gravity.	The Volume becomes after mixing.
100	0	0.7932		66	34	0.8934	96.651	32	68	0.9621	97.056
99	1	0.7969	99.802	65	35	0.8958	96.626	31	69	0.9632	97.158
98	2	0.8006	99.618	64	36	0.8982	96.602	30	70	0.9643	97.268
97	3	0.8042	99.425	63	37	0.9006	96.580	29	71	0.9654	97.367
96	4	0.8078	89.229	62	38	0.9029	96.559	28	72	0.9665	97.466
95	5	0.8114	99.031	61	39	0.9052	96.539	27	73	0.9676	97.565
94	6	0.8150	98.824	60	40	0.9075	96.520	26	74	0.9688	97.664
93	7	0.8185	98.644	59	41	0.9098	96.501	25	75	0.9700	97.763
92	8	0.8219	98.484	58	42	0.9121	96.484	24	76	0.9712	97.862
91	9	0.8253	98.334	57	43	0.9145	96.463	23	77	0.9723	97.958
90	10	0.8286	98.224	56	44	0.9168	96.445	22	78	0.9734	98.051
89	11	0.8317	98.131	55	45	0.9191	96.427	21	79	0.9745	98.149
88	12	0.8346	98.044	54	46	0.9214	96.413	20	80	0.9756	98.262
87	13	0.8373	97.962	53	47	0.9237	96.402	19	81	0.9766	98.377
86	14	0.8400	97.883	52	48	0.9259	96.393	18	82	0.9775	98.494
85	15	0.8427	97.807	51	49	0.9281	96.384	17	83	0.9784	98.613
84	16	0.8454	97.733	50	50	0.9303	96.377	16	84	0.9793	98.731
83	17	0.8481	97.661	49	51	0.9324	96.384	15	85	0.9803	98.845
82	18	0.8508	97.592	48	52	0.9344	96.394	14	86	0.9813	98.955
81	19	0.8534	97.525	47	53	0.9364	96.407	13	87	0.9823	99.058
80	20	0.8566	97.462	46	54	0.9384	96.423	12	88	0.9834	99.154
79	21	0.8591	97.401	45	55	0.9404	96.442	11	89	0.9846	99.246
78	22	0.8616	97.347	44	56	0.9424	96.465	10	90	0.9859	99.333
77	23	0.8642	97.291	43	57	0.9443	96.495	9	91	0.9873	99.413
76	24	0.8668	97.234	42	58	0.9461	96.528	8	92	0.9888	99.487
75	25	0.8695	97.176	41	59	0.9478	96.565	7	93	0.9901	99.555
74	26	0.8723	97.111	40	60	0.9495	96.607	6	94	0.9915	99.617
73	27	0.8751	97.040	39	61	0.9512	96.649	5	95	0.9929	99.674
72	28	0.8779	96.966	38	62	0.9529	96.692	4	96	0.9943	99.731
71	29	0.8806	96.892	37	63	0.9547	96.736	3	97	0.9957	99.792
70	30	0.8833	96.821	36	64	0.9564	96.782	2	98	0.8971	99.857
69	31	0.8860	96.765	35	65	0.9580	96.829	1	99	0.9985	99.927
68	32	0.8885	96.723	34	66	0.9595	96.889	0	100	1.0000	
67	33	0.8910	96.685	33	67	0.9609	96.967				

TABLE

FOR THE MIXING OF ALCOHOL AND WATER BY WEIGHT AT 17°.5 C.

MEISSNER'S Aräometrie.

Parts Alcohol.	Parts Water.	Specific Gravity.	Parts Alcohol.	Parts Water.	Specific Gravity.	Parts Alcohol.	Parts Water.	Specific Gravity.
100	0	0.7932	66	34	0.8806	32	68	0.9543
99	1	0.7960	65	35	0.8831	31	69	0.9561
98	2	0.7988	64	36	0.8855	30	70	0.9578
97	3	0.8016	63	37	0.8879	29	71	0.9594
96	4	0.8045	62	38	0.8902	28	72	0.9608
95	5	0.8074	61	39	0.8925	27	73	0.9621
94	6	0.8104	60	40	0.8948	26	74	0.9634
93	7	0.8135	59	41	0.8971	25	75	0.9647
92	8	0.8166	58	42	0.8994	24	76	0.9660
91	9	0.8196	57	43	0.9016	23	77	0.9673
90	10	0.8225	56	44	0.9038	22	78	0.9686
89	11	0.8252	55	45	0.9060	21	79	0.9699
88	12	0.8279	54	46	0.9082	20	80	0.9712
87	13	0.8304	53	47	0.9104	19	81	0.9725
86	14	0.8329	52	48	0.9127	18	82	0.9738
85	15	0.8353	51	49	0.9150	17	83	0.9751
84	16	0.8376	50	50	0.9173	16	84	0.9763
83	17	0.8399	49	51	0.9196	15	85	0.9775
82	18	0.8422	48	52	0.9219	14	86	0.9786
81	19	0.8446	47	53	0.9242	13	87	0.9796
80	20	0.8470	46	54	0.9264	12	88	0.9806
79	21	0.8494	45	55	0.9280	11	89	0.9817
78	22	0.8519	44	56	0.9308	10	90	0.9830
77	23	0.8543	43	57	0.9329	9	91	0.9844
76	24	0.8567	42	58	0.9350	8	92	0.9860
75	25	0.8590	41	59	0.9371	7	93	0.9878
74	26	0.8613	40	60	0.9391	6	94	0.9897
73	27	0.8635	39	61	0.9410	5	95	0.9914
72	28	0.8657	38	62	0.9429	4	96	0.9931
71	29	0.8680	37	63	0.9448	3	97	0.9948
70	30	0.8704	36	64	0.9467	2	98	0.9965
69	31	0.8729	35	65	0.9486	1	99	0.9982
68	32	0.8755	34	66	0.9505	0	100	1.0000
67	33	0.8781	33	67	0.9524			

TABLE

OF THE PER CENT OF COMMON SALT (NaCl) IN SOLUTIONS OF 1.001
TO 1.20467 SPECIFIC GRAVITY, AT 18°.75 C.

MUSPRATT'S Chemie.

Specific Gravity.	Per cent.	Specific Gravity.	Per cent.	Specific Gravity.	Per cent.
1.001	0.1361	1.034	4.7038	1.067	9.2055
1.002	0.2731	1.035	4.8417	1.068	9.3402
1.003	0.4105	1.036	4.9795	1.069	9.4747
1.004	0.5483	1.037	5.1172	1.070	9.6091
1.005	0.6863	1.038	5.2549	1.071	9.7434
1.006	0.8245	1.039	5.3925	1.072	9.8776
1.007	0.9628	1.040	5.5300	1.073	10.0117
1.008	1.1013	1.041	5.6674	1.074	10.1457
1.009	1.2398	1.042	5.8047	1.075	10.2795
1.010	1.3785	1.043	5.9420	1.076	10.4132
1.011	1.5172	1.044	6.0791	1.077	10.5469
1.012	1.6559	1.045	6.2161	1.078	10.6804
1.013	1.7947	1.046	6.3531	1.079	10.8137
1.014	1.9335	1.047	6.4900	1.080	10.9470
1.015	2.0723	1.048	6.6267	1.081	11.0802
1.016	2.2111	1.049	6.7634	1.082	11.2132
1.017	2.3499	1.050	6.9000	1.083	11.3461
1.018	2.4887	1.051	7.0364	1.084	11.4789
1.019	2.6275	1.052	7.1728	1.085	11.6116
1.020	2.7662	1.053	7.3090	1.086	11.7441
1.021	2.9049	1.054	7.4452	1.087	11.8766
1.022	3.0436	1.055	7.5812	1.088	12.0089
1.023	3.1823	1.056	7.7172	1.089	12.1411
1.024	3.3209	1.057	7.8530	1.090	12.2732
1.025	3.4594	1.058	7.9888	1.091	12.4052
1.026	3.5979	1.059	8.1244	1.092	12.5370
1.027	3.7364	1.060	8.2599	1.093	12.6687
1.028	3.8748	1.061	8.3953	1.094	12.8004
1.029	4.0131	1.062	8.5307	1.095	12.9318
1.030	4.1514	1.063	8.6659	1.096	13.0632
1.031	4.2896	1.064	8.8009	1.097	13.1945
1.032	4.4277	1.065	8.9359	1.098	13.3256
1.033	4.5658	1.066	9.0708	1.099	13.4566

Specific Gravity.	Per cent.	Specific Gravity.	Per cent.	Specific Gravity.	Per cent.
1.100	13.5875	1.136	18.2189	1.171	22.5726
1.101	13.7183	1.137	18.3454	1.172	22.6949
1.102	13.8489	1.138	18.4717	1.173	22.8170
1.103	13.9794	1.139	18.5978	1.174	22.9390
1.104	14.1098	1.140	18.7239	1.175	23.0610
1.105	14.2401	1.141	18.8498	1.176	23.1828
1.106	14.3703	1.142	18.9757	1.177	23.3045
1.107	14.5003	1.143	19.1014	1.178	23.4260
1.108	14.6302	1.144	19.2269	1.179	23.5475
1.109	14.7600	1.145	19.3524	1.180	23.6688
1.110	14.8897	1.146	19.4777	1.181	23.7900
1.111	15.0193	1.147	19.6029	1.182	23.9111
1.112	15.1487	1.148	19.7280	1.183	24.0321
1.113	15.2780	1.149	19.8530	1.184	24.1530
1.114	15.4072	1.150	19.9779	1.185	24.2738
1.115	15.5363	1.151	20.1026	1.186	24.3945
1.116	15.6653	1.152	20.2272	1.187	24.5150
1.117	15.7941	1.153	20.3517	1.188	24.6354
1.118	15.9228	1.154	20.4761	1.189	24.7557
1.119	16.0514	1.155	20.6004	1.190	24.8759
1.120	16.1799	1.156	20.7245	1.191	24.9960
1.121	16.3082	1.157	20.8486	1.192	25.1159
1.122	16.4365	1.158	20.9725	1.193	25.2358
1.123	16.5645	1.159	21.0963	1.194	25.3555
1.124	16.6925	1.160	21.2199	1.195	25.4752
1.125	16.8204	1.161	21.3435	1.196	25.5947
1.126	16.9482	1.162	21.4669	1.197	25.7141
1.127	17.0758	1.163	21.5903	1.198	25.8333
1.128	17.2033	1.164	21.7135	1.199	25.9525
1.129	17.3307	1.165	21.8365	1.200	26.0716
1.130	17.4579	1.166	21.9595	1.201	26.1905
1.131	17.5851	1.167	22.0824	1.202	26.3094
1.132	17.7121	1.168	22.2051	1.203	26.4281
1.133	17.8390	1.169	22.3277	1.204	26.5467
1.134	17.9657	1.170	22.4502	1.20467	26.6261
1.135	18.0924				

TABLES

RELATING TO HEAT.

TABLE

OF THE SPECIFIC AND ATOMIC HEAT OF THE ELEMENTS.

Calculated from Regnault's data.

Name of Element.	Specific Heat.	Atomic Weight.	Specific Heat × Atomic Weight.	Name of Element.	Specific Heat.	Atomic Weight.	Specific Heat × Atomic Weight.
Diamond	0.14680	48.0 ?	6.04640	Mercury, Solid	0.03192	200.0	6.38400
Graphite	0.20180	33.0 ?	6.65940	" Liquid	0.03332	200.0	6.66400
Wood, Charcoal	0.24150			Platinum	0.03243	198.0	6.42114
Silicon, fused	0.17500	35.0 ?	6.12500	Palladium	0.05927	106.4	6.30630
" Crystallized	0.17670			Rhodium	0.05803	104.0	6.03510
Boron	0.25000			Osmium	0.03063	194.0	5.94220
Sulphur (Native)	0.17760	32.0	5.68320	Iridium	0.03259	196.0	6.38760
Selenium	0.08370	79.0	6.61230	Iodine	0.05412	127.0	6.87320
Tellurium	0.04737	128.0	6.06330	Bromine, Solid	0.08430	80.0	6.74400
Magnesium	0.24990	24.0	5.99760	" Liquid	0.10600	80.0	8.48000
Zinc	0.09555	65.0	6.25880	Potassium	0.16956	39.0	6.61280
Cadmium	0.05669	112.0	6.34820	Sodium	0.29340	23.0	6.74800
Aluminum	0.21430	27.5	5.87300	Lithium	0.94080	7.0	6.58560
Iron	0.11379	56.0	6.37220	Phosphorus	0.18870	31.0	5.84970
Nickel	0.10863	58.0	6.30054	Arsenic	0.08140	75.0	6.10500
Cobalt	0.10696	60.0	6.41760	Antimony	0.05077	120.0	6.09240
Manganese	0.12170	54.0	6.57180	Bismuth	0.03084	208.0	6.41472
Tin	0.05623	116.0	6.52268	Thallium	0.03355	204.0	6.84420
Tungsten	0.03342	184.0	6.14920	Silver	0.05701	108.0	6.15700
Copper	0.09515	63.5	6.04190	Gold	0.03244	197.0	6.39068
Lead	0.03140	207.0	6.49990				

TABLE

OF SPECIFIC HEATS OF SOME OF THE MOST COMMON SUBSTANCES.

REGNAULT. — Annales de Chimie et de Physique, III., I., p. 172.

Name.	Specific Heat.	Name.	Specific Heat.
Protoxide of Lead, powder.	0.05118	Bisulphide of Iron (Pyrites)	0.13009
" fused	0 05089	" Tin	0.11932
Oxide of Mercury (HgO)	0.05179	" Molybdenum	0.12334
Protoxide of Manganese	0.15701	Subsulphide of Copper	0.12118
Oxide of Copper	0.14201	" Silver	0.07460
" Nickel	0.16234	Magnetic Pyrites	0.16023
" " calcined	0.15885	Chloride of Sodium	0.21401
Magnesia	0.24394	Chloride of Potassium	0.17295
Oxide of Zinc	0.12480	Subchloride of Mercury	0.05205
Sesquioxide of Iron	0.16695	" Copper	0.13827
Arsenous Acid	0.12786	Chloride of Silver	0.09109
Sesquioxide of Chromium	0.17960	" Barium	0.08570
Oxide of Bismuth	0.06053	" Strontium	0.11990
Teroxide of Antimony	0.09009	" Calcium	0.16420
Corundum	0.19762	" Magnesium	0.19460
Sapphire	0.21732	" Lead	0.06641
Stannic Acid	0.09382	" Mercury	0.06889
Titanic Acid, artificial	0.17164	" Zinc	0.13618
" rutile	0.17032	" Tin	0.10161
Tungstic Acid	0.07983	" Manganese	0.14255
Molybdic Acid	0.13240	Bichloride of Tin	0.14759
Silica	0.19132	" Titanium	0.19145
Boric Acid	0.23743	Terchloride of Arsenic	0.17478
Magnetic Oxide of Iron	0.16780	" Phosphorus	0.20922
Protosulphide of Iron	0.13570	Bromide of Potassium	0.11322
Sulphide of Nickel	0.12814	" Silver	0.07391
" Cobalt	0.12512	" Sodium	0.13842
" Zinc	0.12303	" Lead	0.05326
" Lead	0.05086	Iodide of Potassium	0.08191
" Mercury	0.05117	" Sodium	0.08684
" Tin	0.08365	Subiodide of Mercury	0.03949
" Antimony	0.08403	Iodide of Silver	0.06159
" Bismuth	0.06002	Subiodide of Copper	0.06869

Name.	Specific Heat.	Name.	Specific Heat.
Iodide of Lead	0.04267	Sulphate of Magnesium	0.22159
" Mercury	0.04197	Chromate of Potassium	0.18505
Fluoride of Calcium	0.21492	Bichromate of Potassium	0 18937
Nitrate of Potassium	0.23875	Biborate "	0.21975
" Sodium	0.27821	" Sodium	0.23823
" Silver	0.14352	" Lead	0.11409
" Barium	0.15228	Borate of Potassium	0.20478
Chlorate of Potassium	0.10956	" Sodium	0.25709
Pyrophosphate of Potassium	0.19102	" Lead	0.09046
" Sodium	0.22833	Wolfram	0.09780
" Lead	0.08208	Zircon	0.14558
Metaphosphate of Calcium	0.19923	Carbonate of Potassium	0.21623
Phosphate of Lead	0.07982	" Sodium	0.27275
Arseniate of Potassium	0.15631	Iceland Spar	0.20858
" Lead	0.07280	Arragonite	0.20850
Sulphate of Potassium	0.19010	White Granular Marble	0.21585
" Sodium	0.23115	Gray " "	0.20989
" Barium	0.11285	White Chalk	0.21401
" Strontium	0.14279	Carbonate of Barium	0.11038
" Lead	0.08723	" Iron	0.19345
" Calcium	0.19656	" Strontium	0.14483

LATENT HEAT OF VAPORIZATION.

* FAVRE AND SILBERMAN. — Ann. Chem. Phys. [3] XXXVII. p. 464-470.

† ANDREWS. — Chem. Soc. Qu. Jour. I. p. 27.

Substance.	Latent heat of Vapor.	Substance.	Latent heat of Vapor.
Bromine	45.60†	Butyric Acid	114.67*
Terchloride of Phosphorus	51.42†	Acetate of Ethyl	105.80*
Sulphide of Carbon	86.67†	" "	92.68†
Bichloride of Tin	3.05†	Acetate of Methyl	110.20†
Alcohol	208.92*	Formate of Ethyl	105.30†
"	202.40†	" Methyl	117.10†
Methylic Alcohol	263.86*	Iodide of Ethyl	46.87†
" "	263.70†	" Methyl	46.07†
Amylic Alcohol	121.37*	Oxalate of Ethyl	72.72†
Acetic Acid	101.91*	Butyrate of Methyl	87.33*
Formic Acid	120.72*	Ethal	58.44*
Ether	91.11*	Oil of Turpentine	68.73*
"	90.45†	Terebene	67.21*
Amylic Ether	69.40*	Oil of Lemons	70.02*
Valerianic Acid	103.52*		

TABLE

OF THE SPECIFIC HEAT OF BODIES IN THE SOLID, LIQUID, AND GASEOUS STATES.

REGNAULT.—Mémoires de l'Académie de France, Tome XXVI. p. 227.

Name.	Solid.		Liquid.		Gaseous.	
	Temperature.	Sp. Heat.	Temperature.	Sp. Heat.	Temperature.	Sp. Heat.
	°C.		°C.		°C.	
Water	−78 to 0	0.4740	10	1.0000	128 to 220	0.4805
	−20 to 0	0.5040	10 to 100	1.0000		
Bromine	−77.8 to −25	0.0833	+7.3 to +10	0.1060	83 to 228	0.0555
			+13 to +58	0.1129		
Alcohol			+20	0.5053	105 to 220	0.4584
			0	0.5475		
			+20	0.5951		
			40	0.6479		
			60	0.7060		
			80	0.7694		
Ether			−30	0.5113	70 to 220	0.4797
			0	0.5290		
			+30	0.5467		
			35	0.5497		
Bisulphide of Carbon			−30	0.2303	73 to 192	0.1570
			0	0.2352		
			+30	0.2401		
			45	0.2426		
Wood Spirits			0 to 20	0.6700	101 to 223	0.4580
			−30	0.4824	129 to 233	0.4125
Acetone			0	0.5064		
			+30	0.5302		
			60	0.5540		

Substance	Temperature (°)	Specific heat	Temp. range	Value
Sulphide of Ethyl	20 to 70	0.4785	120 to 223	0.4008
Chloride of Ethyl	−27.6 to +4.5	0.4276	19 to 172	0.2738
Bromide of Ethyl	0 to 20	0.2160	77.7 to 196.5	0.1896
Cyanide of Ethyl	−30	0.4325	114 to 221	0.4262
	0	0.5086		
	+30	0.5847		
	60	0.6608		
	90	0.7369		
Acetate of Ethyl	−30	0.4960	115 to 219	0.4008
	0	0.5274		
	+30	0.5588		
	60	0.5902		
Chloroform	−30	0.2293	117 to 228	0.1567
	0	0.2324		
	+30	0.2354		
	60	0.2384		
Dutch Liquid	−30	0.2790	111 to 221	0.2293
	0	0.2922		
	+30	0.3054		
	60	0.3186		
Benzine	20 to 71	0.4360	116 to 218	0.3754
Turpentine	0	0.4106	179 to 249	0.5061
	40	0.4538		
	80	0.4842		
	120	0.5019		
	160	0.5068		
Chloride Silicon	0 to 20	0.1900	90 to 234	0.1322
Perchloride of Phosphorus	12 to 98	0.2092	112 to 246	0.1347
Terchloride of Arsenic	14 to 98	0.1760	159 to 268	0.1122
Bichloride of Tin	14 to 98	0.1475	149 to 273	0.0939
Bichloride of Titanium	12 to 98	0.1880	162 to 272	0.1290

TABLE

OF SPECIFIC HEATS OF THE MOST COMMON GASES AND VAPORS.

*The quantity of heat necessary to raise 1 gramme of water from 0° to 1°
being unity.*

REGNAULT. — Mémoires de l'Académie de France, Tome XXVI. pp. 303, 313, 318.

Name.	Equal Weights.	Equal Volumes.	Density.
Oxygen	0.21751	0 24049	1.10560
Nitrogen	0.24380	0.23680 ·	0.97130
Hydrogen	3.40900	0.23590	0.06920
Chlorine	0.12099	0.29645	2.45020
Bromine vapor	0.05552	0.30400	5.47720
Deutoxide of Nitrogen	0.23170	0.24060	1.03840
Carbonic oxide	0.24500	0.23700	0.96730
Hydrochloric Acid	0.18520	0.23330	1.25960
Carbonic Acid	0.21690	0.33070	1.52010
Protoxide of Nitrogen	0.22620	0.34470	1.52410
Vapor of Water	0.48050	0.29890	0.62190
Sulphurous Acid	0.15440	0.34140	2.21130
Hydrosulphuric Acid	0.24320	0.28570	1.17470
Bisulphide of Carbon	0.15690	0.41220	2.62580
Protocarburetted Hydrogen	0.59290	0.32770	0.55270
Chloroform	0.15670	0.64610	4.12440
Bicarburetted Hydrogen	0.40400	0.41600	0.96720
Ammonia	0.50840	0.29960	0.58940
Benzine	0.37540	1.01140	2.69420
Essence of Turpentine	0.50610	2.37760	4.69780
Wood Spirit	0.45800	0.50630	1.10550
Ether	0.47970	1.22660	2.55730
Sulphide of Ethyl	0.40080	1.24660	3.11010
Chloride "	0.27380	0.60960	2.22690
Bromide "	0.18960	0.70260	3.70580
Dutch liquid	0.22930	0.78360	3.41740
Acetone	0.41250	0.82640	2.00360
Acetate of Ethyl	0.40080	1.21840	3.04000
Chloride of Silicium	0.13220	0.77780	5.88330
Terchloride of Phosphorus	0.13470	0.63950	4.74640
" Arsenic	0.11220	0.70340	6.26670
Bichloride of Titanium	0.12900	0.85640	6.64020
" Tin	0.09390	0.84160	8.96540

SPECIFIC HEAT OF WATER AND LATENT HEAT OF THE VAPOR OF WATER.

REGNAULT. — Mémoires de l'Académie de France, Tome XXI. p. 748.

Temperature measured by the Air Thermometer. T.	Number of units of Heat lost by one Kilogramme of Water in passing from T° to 0°. Q.	Mean Specific Heat of Water between 0° and T°.	Specific Heat of Water from T° to T°+d T° $\frac{dQ}{dT}$	Latent Heat of the saturated Vapor at the Temperature T.	Total Heat.
0	0.000	0.0000	1.0000	606.5	0.0
10	10.002	1.0002	1.0005	599.5	609.5
20	20.010	1.0005	1.0012	592.6	612.6
30	30.026	1.0009	1.0020	585.7	615.7
40	40.051	1.0013	1.0030	578.7	618.7
50	50.087	1.0017	1.0042	571.6	621.7
60	60.137	1.0023	1.0056	564.7	624.8
70	70.210	1.0030	1.0072	557.6	627.8
80	80.282	1.0035	1.0089	550.6	630.9
90	90.381	1.0042	1.0109	543.5	633.9
100	100.500	1.0050	1.0130	536.5	637.0
110	110.641	1.0058	1.0153	529.4	640.0
120	120.806	1.0067	1.0177	522.3	643.1
130	130.997	1.0076	1.0204	515.1	646.1
140	141.215	1.0087	1.0232	508.0	649.2
150	151.462	1.0097	1.0262	500.7	652.2
160	161.741	1.0109	1.0294	493.6	655.3
170	172.052	1.0121	1.0328	486.2	658.3
180	182.398	1.0133	1.0364	479.0	661.4
190	192.779	1.0146	1.0401	471.6	664.4
200	203.200	1.0160	1.0440	464.3	667.5
210	213.660	1.0174	1.0481	456.8	670.5
220	224.162	1.0189	1.0524	449.4	673.6
230	234.708	1.0204	1.0568	441.9	676.6

TOTAL HEAT OF VAPORIZATION OF DIFFERENT LIQUIDS.

REGNAULT. — Mémoires de l'Académie de France, Tome XXVI. p. 913.

Substance.	°C.	Substance.	°C.
Amylic Alcohol	211.78	Acetate of Ethyl	154.49
Turpentine	139.15	Bisulphide of Carbon	96.80
Essence of Citron	160.49	Alcohol	265.50
Petroleum	194.87	Ether	109.12
Chloride of Ethyl	97.70	Benzine	127.40
Iodide of Ethyl	58.95	Chloroform	76.33
Bromine	50.95	Chloride of Carbon (CCl_2)	62.30
Bichloride of tin	46.84	Acetone	153.65
Terchloride of Arsenic	69.74	Water	636.70
Terchloride of Phosphorus	67.24		

TABLE

Substance.	Expansion.	Observer.
Lead	0.00284836	Lavoisier.
"	0.00271900	Guyton de Morveau.
"	0.00286667	Smeaton.
"	0.00300500	Calvert and Johnson.
Iron, Wrought	0.00122045	Lavoisier and Laplace.
" "	0.00118203	Dulong and Petit.
" "	0.00125833	Smeaton.
" "	0.00118700	Calvert and Johnson.
Iron, Wire	0.00123504	Lavoisier.
" "	0.00114010	Borda.
Iron, Cast	0.00111700	Calvert and Johnson.
" "	0.00110940	Roy.
Glass, English flint	0.00081166	· Lavoisier.
" French "	0.00087199	"
" Soft	0.00086100	Dulong and Petit.
" "	0.00083333	Smeaton.
" Tube	0.00077615	Roy.
" " free from Lead	0.00087572	Lavoisier.
" " " " "	0.00091751	"
" Rod	0.00080787	Roy.
Gold, Fine	0.00146606	Lavoisier.
" Paris test unannealed	0.00155155	"
" " " annealed	0.00151361	"
"	0.00147500	Guyton de Morveau.
"	0.00140100	Ellicot.
"	0.00137400	Calvert and Johnson.
Copper	0.00171000	Ellicot.
" Hammered	0.00172244	Lavoisier.
" "	0.00176900	Calvert and Johnson.
"	0.00171222	Lavoisier.
"	0.00171822	Dulong and Petit.
" Cast	0.00187900	Calvert and Johnson.
Brass	0.00193332	Smeaton.

Substance.	Expansion.	Observer.
Brass Wire	0.00187821	Lavoisier.
" "	0.00188500	Herbert.
" Hammered	0.00182800	Calvert and Johnson.
" Cast	0.00186671	Lavoisier.
" "	0.00193000	Calvert and Johnson.
"	0.00187500	Smeaton.
Palladium	0.00100000	Wollaston.
Platinum	0.00088417	Dulong and Petit.
"	0.00090000	Wollaston.
"	0.00085700	Guyton de Morveau.
"	0.00088100	Calvert and Johnson.
Silver, Cupeled	0.00190974	Lavoisier.
"	0.00198800	Guyton de Morveau.
"	0.00208260	Troughton.
" Cast	0.00199100	Calvert and Johnson.
Steel, Soft	0.00107875	Lavoisier.
" "	0.00115000	Smeaton.
" "	0.00103800	Calvert and Johnson.
" Tempered	0.00140200	" "
" "	0.00136900	Lavoisier.
" Rod	0.00114470	Roy.
Bismuth	0.00139167	Smeaton.
"	0.00134100	Calvert and Johnson.
Zinc, Cast	0.00294167	Smeaton.
" "	0.00305100	Guyton de Morveau.
" Hammered	0.00310833	Smeaton.
" "	0.00219300	Calvert and Johnson.
Tin, Fine	0.00216400	Guyton de Morveau.
"	0.00209300	Herbert.
"	0.00271700	Calvert and Johnson.
" Malacca	0.00193765	Lavoisier.
" Common	0.00248330	Smeaton.
Bronze	0.00181667	"
Alloy 1 Zn 2 Cu	0.00205833	"
Calc-spar in the Chief Axis	0.00286000	Mitscherlich.
Coal from Oak Wood	0.00120000	P. Henrich.
" " Fir "	0.00100000	"
Marble from Carrara	0.00084870	Destigny.
" " St. Beat	0.00041810	"
Phosphorus from 0 – 39.50°	0.00142455	Erman.
Aluminum	0.00221800	Calvert and Johnson.
Cadmium	0.00332300	" "
Antimony, Hammered	0.00098500	" "
Speculum Metal	0.00193333	Smeaton.
Antimony	0.00108330	"
Solder, Soft	0.00250533	"

TABLE

OF THE CUBICAL EXPANSION OF DIFFERENT SOLID BODIES FOR EACH
INCREASE OF 1° C. OF TEMPERATURE.

KOPP. — Annalen der Chemie und Pharmacie, LXXXL 65.

Name of Substance.	Formula.	Coefficient of Expansion.
Antimony	Sb	0.000033
Bismuth	Bi	0.000040
Cadmium	Cd	0.000094
Copper	Cu	0.000051
Iron	Fe	0.000037
Lead	Pb	0.000089
Sulphur	S	0.000183
Tin	Sn	0.000069
Zinc	Zn	0.000089
Barium, Sulphate	BaO, SO_3	0.000058
Arragonite	CaO, CO_2	0.000065
Dolomite	$CaO, CO_2 + MgO, CO_2$	0.000035
Strontium, Sulphate	SrO, SO_3	0.000061
Specular Iron Ore	Fe_2O_3	0.000040
Iron Pyrites	FeS_2	0.000034
Carbonate of Iron, Manganese, and Magnesia	$Fe(Mg, Mn)O, CO_2$	0.000035
Fluor Spar	CaF	0.000062
Carbonate of Lime, Calc-spar	CaO, CO_2	0.000018
Magnetic Iron Ore	Fe_3O_4	0.000029
Orthoclase	$\begin{cases} KO, SiO_2 + \\ Al_2O_3, 3SiO_2 \end{cases}$	0.000026 0.000017
Quartz	SiO_2	$\begin{cases} 0.000042 \\ 0.000039 \end{cases}$
Rutile	TiO_2	0.000032
Tin Stone	SnO_2	0.000016
Zinc Blende	ZnS	0.000036
Hard Potash Glass		0.000021
Soft Soda Glass		0.000026
" " " another kind		0.000024
Galena	PbS	0.000068

TABLE

OF THE VOLUME OF A GLASS VESSEL WHEN ITS VOLUME AT 15° C.
IS TAKEN AS UNITY.

GERLACH'S Salzlösungen, p. 115.

VOLUME AT			
°C.		°C.	
0	0.99961210	23	1.00020688
1	0.99963796	24	1.00023274
2	0.99966382	25	1.00025860
3	0.99968968	26	1.00028446
4	0.99971554	27	1.00031032
5	0.99974140	28	1.00033618
6	0.99976726	29	1.00036204
7	0.99979313	30	1.00038790
8	0.99981898	35	1.00051720
9	0.99984484	40	1.00064650
10	0.99987070	45	1.00077580
11	0.99989656	50	1.00090510
12	0.99992242	55	1.00103440
13	0.99994828	60	1.00116370
14	0.99997414	65	1.00129300
15	1.00000000	70	1.00142230
16	1.00002586	75	1.00155160
17	1.00005172	80	1.00168090
18	1.00007758	85	1.00181020
19	1.00010344	90	1.00193950
20	1.00012930	95	1.00206880
21	1.00015516	100	1.00219810
22	1.00018102		

TABLE

OF THE EXPANSION OF GLASS.

Calculated from Regnault's data.

JAS T. BROWN.—Jour. Chem. Society, N. S. IV. p. 80.

$t' - t$	$1 + k\,(t' - t)$	$t' - t$	$1 + k\,(t' - t)$	$t' - t$	$1 + k\,(t' - t)$
35	1.00096	145	1.00421	250	1.00765
40	1.00110	150	1.00436	255	1.00780
45	1.00124	155	1.00451	260	1.00795
50	1.00138	160	1.00465	265	1.00810
55	1.00151	165	1.00480	270	1.00826
60	1.00165	170	1.00494	275	1.00841
65	1.00179	175	1.00509	280	1.00856
70	1.00193	180	1.00523	285	1.00892
75	1.00207	185	1.00551	290	1.00907
80	1.00220	190	1.00566	295	1.00923
85	1.00241	195	1.00581	300	1.00939
90	1.00255	200	1.00596	305	1.00954
95	1.00269	205	1.00610	310	1.00970
100	1.00284	210	1.00625	315	1.00985
105	1.00298	215	1.00640	320	1.01001
110	1.00312	220	1.00655	325	1.01017
115	1.00326	225	1.00670	330	1.01032
120	1.00340	230	1.00685	335	1.01048
125	1.00355	235	1.00719	340	1.01064
130	1.00369	240	1.00734	345	1.01079
135	1.00392	245	1.00749	350	1.01095
140	1.00407				

TABLE

OF THE ABSOLUTE DILATATION OF MERCURY.

REGNAULT. — Mémoires de l'Académie de France, Tome XXI. p. 328.

Temperature by Air Thermometer. $T.$	Dilatation of Mercury from $0°$ to $T.$ δ_r	Mean Coefficient of Dilatation from $0°$ to $T.$ δ	Real Coefficient of Dilatation at $T.$ $\dfrac{d\,\delta_r}{d\,T}$	Temperature deduced from the absolute Dilatation of Mercury. θ	Difference between Thermometer based on the absolute Dilatation of Mercury and the Air Thermometer. $(\theta - T.)$
0	0.000000	0.00000000	0.00017905	0.000	0.000
10	0.001792	0 00017925	0.00017950	9.872	—0.128
20	0.003590	0.00017951	0.00018001	19.776	—0.224
30	0.005393	0.00017976	0.00018051	29.709	—0.291
40	0.007201	0.00018002	0.00018102	39.668	—0.332
50	0.009013	0.00018027	0.00018152	49.650	—0.350
60	0.010831	0.00018052	0.00018203	59.665	—0.335
70	0.012655	0.00018078	0.00018253	69.713	—0.287
80	0.014482	0.00018102	0.00018304	79.777	—0.223
90	0.016315	0.00018128	0.00018354	89.875	—0.125
100	0.018153	0.00018153	0.00018405	100.000	0.000
110	0.019996	0.00018178	0.00018455	110.153	+0.153
120	0.021844	0.00018203	0.00018505	120.333	0.333
130	0.023697	0.00018228	0.00018556	130.540	0.540
140	0.025555	0.00018254	0.00018606	140.776	0.776
150	0.027419	0.00018279	0.00018657	151.044	1.044
160	0.029287	0.00018304	0.00018707	161.334	1.334
170	0.031160	0.00018329	0.00018758	171.652	1.652
180	0.033039	0.00018355	0.00018808	182.003	2.003
190	0.034922	0.00018380	0.00018859	192.376	2.376
200	0.036811	0.00018405	0.00018909	202.782	2.782
210	0.038704	0.00018430	0.00018959	213.210	3.210
220	0.040603	0.00018456	0.00019010	223.671	3.671
230	0.042506	0.00018481	0.00019061	234.154	4.154
240	0.044415	0.00018506	0.00019111	244.670	4.670
250	0.046329	0.00018531	0.00019161	255.214	5.214
260	0.048247	0.00018557	0.00019212	265.780	5.780
270	0.050171	0.00018582	0.00019262	276.379	6.379
280	0.052100	0.00018607	0.00019313	287.005	7.005
290	0.054034	0.00018632	0.00019363	297.659	7.659
300	0.055973	0.00018658	0.00019413	308.340	8.340
310	0.057917	0.00018683	0.00019464	319.048	9.048
320	0.059866	0.00018708	0.00019515	329.786	9.786
330	0.061820	0.00018733	0.00019565	340.550	10.550
340	0.063778	0.00018758	0.00019616	351.336	11.336
350	0.065743	0.00018784	0.00019666	362.160	12.160

TABLE

COMPARING AIR AND MERCURY THERMOMETERS MADE OF DIFFER-
ENT KINDS OF GLASS.

REGNAULT. — Mémoires de l'Académie de France, Tome XXI. p. 239.

Temperature of Air Thermometer.	Temperatures of Mercurial Thermometers.			
	Crystal from Choisy-le-Roi.	Common Glass.	Green Glass.	Hard Swedish Glass.
100	100.00	100.00	100.00	100.00
110	110.05	109.98	110.03	110.02
120	120.12	119.95	120.08	120.04
130	130.20	129.91	130.14	130.07
140	140.29	139.85	140.21	140.11
150	150.40	149.80	150.30	150.15
160	160.52	159.74	160.40	160.20
170	170.65	169.68	170.50	170.26
180	180.80	179.63	180.60	180.33
190	191.01	189.65	190.70	190.41
200	201.25	199.70	200.80	200.50
210	211.53	209.75	211.00	210.61
220	221.82	219.80	221.20	220.75
230	232.16	229.85	231.42	230.91
240	242.55	239.90	241.60	241.16
250	253.00	250.05	251.85	251.44
260	263.44	260.20	262.15	
270	273.90	270.38	272.50	
280	284.48	280.52	282.85	
290	295.10	290.80	293.30	
300	305.72	301.08		
310	316.45	311.45		
320	327.25	321.80		
330	338.22	332.40		
340	349.30	343.00		
350	360.50	354.00		

TABLE

FOR COMPARISON OF CENTIGRADE AND FAHRENHEIT THERMOMETER
SCALES.

Centi-grade.	Fahrenheit.	Centi-grade.	Fahrenheit.	Centi-grade.	Fahrenheit.	Centi-grade.	Fahrenheit.
+100	+212.0	+64	+147.2	+29	+84.2	− 6	+21.2
99	210.2	63	145.4	28	82.4	7	19.4
98	208.4	62	143.6	27	80.6	8	17.6
97	206.6	61	141.8	26	78.8	9	15.8
96	204.8	60	140.0	25	77.0	10	14.0
95	203.0	59	138.2	24	75.2	11	12.2
94	201.2	58	136.4	23	73.4	12	10.4
93	199.4	57	134.6	22	71.6	13	8.6
92	197.6	56	132.8	21	69.8	14	6.8
91	195.8	55	131.0	20	68.0	15	5.0
90	194.0	54	129.2	19	66.2	16	3.2
89	192.2	53	127.4	18	64.4	17	+ 1.4
88	190.4	52	125.6	17	62.6	18	− 0.4
87	188.6	51	123.8	16	60.8	19	2.2
86	186.8	50	122.0	15	59.0	20	4.0
85	185.0	49	120.2	14	57.2	21	5.8
84	183.2	48	118.4	13	55.4	22	7.6
83	181.4	47	116.6	12	53.6	23	9.4
82	179.6	46	114.8	11	51.8	24	11.2
81	177.8	45	113.0	10	50.0	25	13.0
80	176.0	44	111.2	9	48.2	26	14.8
79	174.2	43	109.4	8	46.4	27	16.6
78	172.4	42	107.6	7	44.6	28	18.4
77	170.6	41	105.8	6	42.8	29	20.2
76	168.8	40	104.0	5	41.0	30	22.0
75	167.0	39	102.2	4	39.2	31	23.8
74	165.2	38	100.4	3	37.4	32	25.6
73	163.4	37	98.6	2	35.6	33	27.4
72	161.6	36	96.8	+ 1	33.8	34	29.2
71	159.8	35	95.0	0	32.0	35	31.0
70	158.0	34	93.2	− 1	30.2	36	32.8
69	156.2	33	91.4	2	28.4	37	34.6
68	154.4	32	89.6	3	26.6	38	36.4
67	152.6	31	87.8	4	24.8	39	38.2
66	150.8	+30	+ 86.0	− 5	+23.0	−40	−40.0
+65	+149.0						

TABLE

FOR THE COMPARISON OF THE FAHRENHEIT AND CENTIGRADE THERMOMETER SCALES.

Fahrenheit.	Centigrade.	Fahrenheit.	Centigrade.	Fahrenheit.	Centigrade.
+212	+100.00	+174	+78.89	+136	+57.78
211	99.44	173	78.33	135	57.22
210	98.89	172	77.78	134	56.67
209	98.33	171	77.22	133	56.11
208	97.78	170	76.67	132	55.55
207	97.22	169	76.11	131	55.00
206	96.67	168	75.55	130	54.44
205	96.11	167	75.00	129	53.89
204	95.55	166	74.44	128	53.33
203	95.00	165	73.89	127	52.78
202	94.44	164	73.33	126	52.22
201	93.89	163	72.78	125	51.67
200	93.33	162	72.22	124	51.11
199	92.78	161	71.67	123	50.55
198	92.22	160	71.11	122	50.00
197	91.67	159	70.55	121	49.44
196	91.11	158	70.00	120	48.89
195	90.55	157	69.44	119	48.33
194	90.00	156	68.89	118	47.78
193	89.44	155	68.33	117	47.22
192	88.89	154	67.78	116	46.67
191	88.33	153	67.22	115	46.11
190	87.78	152	66.67	114	45.55
189	87.22	151	66.11	113	45.00
188	86.67	150	65.55	112	44.44
187	86.11	149	65.00	111	43.89
186	85.55	148	64.44	110	43.33
155	85.00	147	63.89	109	42.78
184	84.44	146	63.33	108	42.22
183	83.89	145	62.78	107	41.67
182	83.33	144	62.22	106	41.11
181	82.78	143	61.67	105	40.55
180	82.22	142	61.11	104	40.00
179	81.67	141	60.55	103	39.44
178	81.11	140	60.00	102	38.89
177	80.55	139	59.44	101	38.33
176	80.00	138	58.89	100	37.78
+175	+79.44	+137	+58.33	+99	+37.22

Fahrenheit.	Centigrade.	Fahrenheit.	Centigrade.	Fahrenheit.	Centigrade.
+98	+36.67	+51	+10.55	+ 5	−15.00
97	36.11	50	10.00	4	15.55
96	35.55	49	9.44	3	16.11
95	35.00	48	8.89	2	16.67
94	34.44	47	8.33	+ 1	17.22
93	33.89	46	7.78	0	17.78
92	33.33	45	7.22	− 1	18.33
91	32.78	44	6.67	2	18.89
90	32.22	43	6.11	3	19.44
89	31.67	42	5.55	4	20.00
88	31.11	41	5.00	5	20.55
87	30.55	40	4.44	6	21.11
86	30.00	39	3.89	7	21.67
85	29.44	38	3.33	8	22.22
84	28.89	37	2.78	9	22.78
83	28.33	36	2.22	10	23.33
82	27.78	35	1.67	11	23.89
81	27.22	34	1.11	12	24.44
80	26.67	33	+ 0.55	13	25.00
79	26.11	32	0.00	14	25.55
78	25.55	31	− 0.55	15	26.11
77	25.00	30	1.11	16	26.67
76	24.44	29	1.67	17	27.22
75	23.89	28	2.22	18	27.78
74	23.33	27	2.78	19	28.33
73	22.78	26	3.33	20	28.89
72	22.22	25	3.89	21	29.44
71	21.67	24	4.44	22	30.00
70	21.11	23	5.00	23	30.55
69	20.55	22	5.55	24	31.11
68	20.00	21	6.11	25	31.67
67	19.44	20	6.67	26	32.22
66	18.89	19	7.22	27	32.78
65	18.33	18	7.78	28	33.33
64	17.78	17	8.33	29	33.89
63	17.22	16	8.89	30	34.44
62	16.67	15	9.44	31	35.00
61	16.11	14	10.00	32	35.55
60	15.55	13	10.55	33	36.11
59	15.00	12	11.11	34	36.67
58	14.44	11	11.67	35	37.22
57	13.89	10	12.22	36	37.78
56	13.33	9	12.78	37	38.33
55	12.78	8	13.33	38	38.89
54	12.22	7	13.89	39	39.44
53	11.67	+ 6	−14.44	−40	−40.00
+52	+11.11				

EXPANSIONS AND BOILING-POINTS OF LIQUIDS.

General Formula, $V = 1 \pm At \pm Bt^2 \pm Ct^3$ *in which* $V =$ *the Volume at any Temperature t, the Volume at 0° being unity.*

PIERRE. — Annales de Chemie et de Physique, 3 série, Tomes XV. p. 325, XIX. p. 193, XX. p. 5, XXI. p. 336, XXXI. p. 118, XXXIII p. 199.

Name.	Formula.	Boiling Point at 700 mm.	A	B	C	Specific Gravity at 0°.	Remarks.
Alcohol	$C_2H_6O_2$	78.3	+.001048630	+.017509900	+.000134518	0.81510	
Wood Spirit	$C_2H_4O_2$	66.3	.001185569	.015649300	.000911100	0.82074	
Bisulphide of Carbon	CS_3	48.0	.001139800	.013706500	.001912200	1.29310	
Ether	$C_8H_{10}O_2$	35.7	.001513240	.023590000	.004005100	0.73580	
Chloride of Ethyl	$C_8H_{10}Cl_2$	11.0	.001574578	.028136000	.001569870	0.92116	
Bromide of Ethyl	$C_8H_{10}Br_2$	40.8	.001337600	.015013480	.001690000	1.47329	
Iodide of Ethyl	$C_8H_{10}I_2$	70.3	.001142250	.019638000	.000620640	1.97540	
Bromide of Methyl	$C_2H_6Br_2$	13.0	.001415200	.033152810	.011380900	1.66440	
Iodide of Methyl	$C_2H_6I_2$	41.1	.001199590	.021633180	.001005000	2.19920	
Formate of Ethyl	$C_6H_6O_4$	53.2	.001325200	.028624840	.000661800	0.93560	
Acetate of Methyl	$C_6H_6O_4$	59.5	.001295950	.029098000	.000425690	0.86680	
Fusel Oil	$C_{10}H_{12}O_2$	132.1	.000890100	.006572900	.001184580	0.82705	— 15° to + 80°
			.008938500	.006874460	.001009600		+ 80° to + 182°.1
Acetate of Ethyl	$C_8H_8O_4$	74.0	.001258490	.029568800	.001492150	0.90691	
Butyrate of Methyl	$C_{10}H_{10}O_4$	102.7	.001239890	.006260240	.001306500	1.02928	13° to 99°
" Ethyl	$C_{12}H_{12}O_4$	119.4	.001202790	.000722338	.002263460	0.90190	99° to 119°.4
Chloride of Phosphorus	PCl_3	78.74	.001128600	.008728800	.005027800	1.61616	0° to 100°
Bromide of Phosphorus	PBr_3	175.3	.000847200	.043367000	.000252700	2.92489	100° to 175°.3
Chloride of Arsenic	$AsCl_3$	133.91	.000979072	.009669000	.000177700	2.20495	
Bichloride of Tin	$SnCl_2$	115.7	.001132800	.009117100	.000757900	2.26712	
" Titanium	$TiCl_2$	135.9	+.000942569	.013457900	.000068800	1.76088	

Substance	Formula	δ = t − t'	+.001294100	+.021840000	+.004086400	Density (water at 4° = unity) 1.52370	Water at 4° = unity.
Chloride of Silicon	$SiCl_3$	59.0	.001294100	.021840000	.004086400	1.52370	melt'g pt. 20°.09
Bromide of Silicon	$SiBr_3$	158.26	.000952570	.007567000	.000029200	2.81280	20°.09 to 100°.16
Chloride of Ethylene	$C_2H_2Cl_2$	84.92	.001118900	.010468600	.001034100	1.28034	100°.16 to 132°.8
Bromide of Ethylene	$C_4H_4Br_2$	132.8	.0009526908² *	.0131650008²	.0010162688³	2.16292	
			.0010167608² **	.0010223008²	.0008788008³		
Bromine	Br	63.04	.0010038180	.017113800	.000544700	3.18700	−25°.85 as unity.
Sulphurous Acid	SO_2	−8.0	.0014960008³	.2233700008³	.0495700008³	1.49110	
Sulphite of Ethyl	$C_4H_5SO_3$	160.2	.000993479	.010903880	.000153900	1.10634	0° to 100°
Aldehyde	$C_4H_4O_3$	22.1	.001653500	.085060000	.006425800	0.80551	
Butyric Acid	$C_8H_8O_4$	163.4	.001025700	.008186000	.000346900	0.98165	100° to 163°.4
			.001030400	.008188000	.000333200		
Chloride of Chlorethyl	$C_2H_2Cl_3$	65.0	.001290700	.001183310	.002133900	1.24074	0° to 75°
" Chlorethylene	$C_4H_3Cl_3$	114.4	.001056000	.002803500	.001508800	1.42200	75° to 114°.4
Chloride of Bichlorethyl	$C_2H_3Cl_3$	75.0	.000952700	.031950000	+.000641200	1.34650	
Bisulphide of Methyl	$C_2H_3S_2$	112.7	.001174800	.035770000	−.000536700	1.06358	0° to 70°
Sulphocyanide of Methyl	$C_4H_3NS_2$	132.96	.001017000	.015760000	−.000190700	1.08794	70° to 132°.96
			.000970070	.012540000	+.001175400		
			.000948000	.025479000	−.000246400		
Chloroform	C_2HCl_3	63.0	.001107000	.046647000	−.001743200	1.52520	
Bichloride of Carbon	C_2Cl_4	78.6	.001183400	.008988000	+.001351300	1.62983	
Sulphide of Ethyl	$C_8H_{10}S_2$	91.0	.001196400	.018065000	.000788200	0.83672	
Chloride of Amyl	$C_{10}H_{11}Cl$	102.0	.001171500	.005007700	.001353684	0.89584	0° to 80°
Bromide of Amyl	$C_{10}H_{11}Br$	118.6	.001023000	.019008600	.000197500	1.16576	80° to 118°.6
			.001070900	.008544500	+.000764000		
Chloride of Bichlorethylene	$C_4H_2Cl_4$	138.5	.0008683619	.065877000	−.005411000	1.61158	0° to 60°
			.009771680	.007347800	+.000401000		60° to 138°.5
" Trichlorethylene	C_4HCl_4	153.7	.000899000	.024577700	−.001286400	1.66267	0° to 75°
			.000973390	.000257700	.000636400		75° to 153°.7
Protochloride of Carbon	C_4Cl_4	123.9	.001002600	.003279800	+.001593390	1.64900	0° to 75°
			.000920800	.034007450	−.001007500		75° to 123°.9
Terebene	$C_{20}H_{16}$	161.6	.000896550	.020366800	.000748360	0.87179	0° to 80°
			−.0008879250	+.018095000	−.000172600		80° to 161°.6

* δ = t − t'   t' = the melting point or point taken as unity.

EXPANSIONS AND BOILING-POINTS OF LIQUIDS.

General Formula for the Expansion of Liquids, $V = 1 \pm A t^\circ \pm B t^{\circ 2} \pm C t^{\circ 3}$.

KOPP.— Annalen der Physik u. Chemie, LXXII. 1–223. Annalen der Chem. und Phar. XCIV. 257, XCV. 307, XCVIII. 307.

Name.	Formula.	Boiling Point at 760 mm. °C	A	B	C	Specific Gravity at 0°.	Remarks.
				0.00000	0.000000		
Wool Spirit	$C_2H_4O_2$	65.8	+.00113420	+.13635000	+.00874100	0.81420	
Fusel Oil	$C_{10}H_{12}O_2$	132.2	.00097240	—.08565100	—.02021800	0.82480	
Valeraldehyde	$C_{10}H_{10}O_2$	93.6	.00119360	+.29750000	—.00418070	0.82240	
Anhydrous Acetic Acid	$C_8H_{12}O_3$	137.9	.00105307	+.18389000	+.00079165	1.09690	
Acetate of Amyl	$C_{14}H_{14}O_4$	137.9	.00115010	—.00904600	.01301500	0.88370	
Valerate of Amyl	$C_{20}H_{20}O_4$	188.9	.00103170	+.00832540	.00768980	0.87930	
Oxalate of Ethyl	$C_{12}H_{10}O_8$	186.1	.00106880	.08417000	.00472550	1.10160	
Salicylate of Methyl	$C_{16}H_8O_6$	223.7	.00084360	.04008200	.00255050	1.19690	
Benzoic Acid	$C_{14}H_6O_4$	249.9	.00080370δ*	.124590000δ*		1.08380	melts at 121°.4
Benzoate of Methyl	$C_{16}H_8O_4$	199.7	.00089390	—.08529000	.00259360	1.10260	
" Ethyl	$C_{18}H_{10}O_4$	213.4	.00093094	—.00634290	.00499280	1.06570	
" Amyl	$C_{24}H_{10}O_4$	261.2	.00082495	+.07303500	.00128330	1.00390	
Benzoic Alcohol	$C_{14}H_8O_2$	206.8	.00078730	.05129900	.00272500	1.06280	
Oil of Bitter Almonds	$C_{14}H_6O_2$	179.4	.00094020	—.08204500	.00806000	1.06360	
Cuminol	$C_{20}H_{12}O_2$	237.0	.00084150	+.02222000	.00348430	0.98320	
Cymol	$C_{20}H_{14}$	178.1	.00094060	.03808500	.00486670	0.87780	
Propionic Acid	$C_6H_6O_4$	141.8	.00110030	+.02181600	.00697960	1.01610	
Valeric "	$C_{10}H_{10}O_4$	176.3	.00104760	—.02400100	+.00824660	0.95550	
Phenol	$C_{12}H_6O_2$	188.6	.00067440	+.17210000	—.00050408	1.08080	
Propionate of Ethyl	$C_{10}H_{10}O_4$	97.7	.00128600	.05138600	+.01730500	0.92310	
Cinnamate "	$C_{22}H_{10}O_4$	266.6	.00081090	.06401600	.00143760	1.06560	
Oxalate of Methyl	$C_8H_6O_8$	161.1	.001079008δ*	.1555400008δ*		1.15660	melts at 50°.
Carbonate of Ethyl	$C_{10}H_{10}O_6$	126.4	.00117110	.05259600	.00985210	0.99980	
Succinate "	$C_{16}H_{14}O_8$	217.7	.00100880	.03328200	.00517010	1.07180	

* δ = Number of degrees above melting point.

Substance	Formula						Notes
Naphthalin	$C_{20}H_8$	216.8	+.000747008	+1809500008²	+.01629700	0.97740	melts at 79°.2
Butyl	$C_{16}H_{18}$	109.0	.00121250	02793000		0.71350	
Chloride of Amyl	$C_{10}H_{11}Cl$	101.4	.00092940	31403000	—0492100	0.88590	
" Butyl	$C_8H_{11}Cl$	122.8	.00131540	33706000		1.09530	
" Acetyl	$C_4H_3O_2Cl$	56.0	.00085893	04421900		1.13050	
" Benzoyl	$C_{14}H_3O_2Cl$	198.4	.00095450	22189000	+.00271390	1.23240	
Chloral	$C_4HCl_3O_2$	93.6	.00096500	12314000	05639200	1.51830	
Iodide of Amyl	$C_{10}H_{11}I$	148.4	.00103250	17259000	00241110	1.46760	
Amyl Mercaptan	$C_{10}H_{12}S_2$	120.1			+.00153180	0.85480	
Chloride of Antimony	$SbCl_3$	223.5	.000805408²	103300008²		2.67500	melts at 73°.2
Bromide "	$SbBr_3$	275.4	.000576008²	134650008²		3.64100	" 90°.0
Chloride of Sulphur	S_2Cl	144.0	.00095010	00381850	+.00731860	1.70550	
Nitrate of Ethyl	$C_4H_5NO_6$	87.4	.00112900	47915000	—01841300	1.13220	
Nitrobenzol	$C_{12}H_5NO_4$	220.2–221.1	.00082630	05224900	+.00137790	1.20020	
Aniline	$C_{12}H_7N$	185.6	.00081730	09191000	00062784	1.03610	
Cyanide of Methyl	C_4H_3N	71.4–72.6	.00121180	17780000	01532200	0.83470	
" Phenyl	$C_{14}H_5N$	191.5	.00093380	03072200	00579600	1.02300	
Mustard Oil	$C_8H_5NS_2$	151.8	.00107130	00327010	00735690	1.02820	
Alcohol	$C_4H_6O_2$	78.4	.00101139	07836000	01761800	0.80950	
Ether	$C_8H_{10}O_2$	34.9	.00118026	350531600	02700700	0.73660	
Aldehyde	$C_4H_4O_2$	20.8	.00154640	69745000		0.80092	
Acetone	C_3H_6O	56.3	.00131810	26090000	01056830	0.81440	
Benzol	$C_{12}H_6$	80.8	.00117626	12775500	00806480	0.89911	
Formic Acid	$C_2H_2O_4$	105.3	.00099269	06251400	00596500	1.22270	
Acetic "	$C_4H_4O_4$	117.3	.00105703	01832300	00964350	1.08005	
Butyric "	$C_8H_8O_4$	157.0	.00104610	05624400	00542010	0.98860	
Formate of Methyl	$C_4H_4O_4$	33.4	.00140550	17131000	04594700	0.99840	
" " Ethyl	$C_6H_6O_4$	54.9	.00136446	01358800	03924800	0.94474	
Acetate of Methyl	$C_6H_6O_4$	56.3	.00127790	39471000	00363900	0.95620	
" " Ethyl	$C_6H_6O_4$	74.3	.00127380	21914000	01179700	0.91046	
Butyrate of Methyl	$C_{10}H_{10}O_4$	98.6	.00119565	18103000	00982920	0.92098	
" Ethyl	$C_{12}H_{12}O_4$	114.8	.00117817	13093000	00956000	0.90412	
Valerate of Methyl	$C_{12}H_{12}O_4$	116.2	.00112115	01701400	+.00586270	0.90150	

TABLE

OF THE ELEVATION OF THE BOILING POINT BY DISSOLVING SALTS
IN WATER.

LEGRAND. — Annales de Chemie et de Physique.

The salts are nearly all anhydrous.

Name of Salt.	Boiling Point C°.	Proportion of salt to 100 of Water.
Chloride of Barium	104.40	60.1
Chlorate of Potassium	104.20	61.5
Carbonate of Sodium	104.63	48.5
Phosphate of Sodium	106.60	112.6
Chloride of Potassium	108.30	59.4
Chloride of Sodium	108.40	41.2
Chloride of Ammonium	114.20	88.9
Tartrate of Potassium	114.67	296.2
Nitrate of Potassium	115.90	335.1
Chloride of Strontium	117.85	117.5
Nitrate of Sodium	121.00	224.0
Acetate of Sodium	124.37	209.0
Carbonate of Potassium	135.00	205.0
Nitrate of Calcium	151.00	362.2
Acetate of Potassium	169.00	798.2
Chloride of Lime	179.50	325.0
Nitrate of Ammonia	180.00	Any Amount.

TABLE

OF THE VOLUME OF WATER AT DIFFERENT TEMPERATURES.

Temperature.	Kopp.	Calculated by Frankenheim from Pierre's formulas.	Depretz. Water at 4° C = 1.
—15°		1.0037584	
10		1.0016851	
9		1.0014013	1.0016311
8		1.0011526	1.0013734
7		1.0009355	1.0011354
6		1.0007465	1.0009184
5		1.0005819	1.0006987
4		1.0004382	1.0005619
3		1.0003117	1.0004222
2		1.0001989	1.0003077
— 1		1.0000962	1.0002138

Temperature.	Kopp.	Calculated by Frankenheim from Pierre's formulas.	Depretz. Water at 4° C = 1.
0°	1.000000	1.0000000	1.0001269
+ 1	0.999947	0.9999458	1.0000730
2	0.999908	0.9999094	1.0000331
3	0.999885	0.9998878	1.0000083
4	0.999877	0.9998820	1.0000000
5	0.999883	0.9998903	1.0000082
6	0.999903	0.9999148	1.0000309
7	0.999938	0.9999524	1.0000708
8	0.999986	1.0000044	1.0001216
9	1.000048	1.0000694	1.0001879
10	1.000124	1.0001482	1.0002684
11	1.000213	1.0002392	1.0003598
12	1.000314	1.0003420	1.0004724
13	1.000429	1.0004557	1.0005862
14	1.000556	1.0005877	1.0007146
15	1.000695	1.0007275	1.0008751
16	1.000846	1.0008784	1.0010215
17	1.001010	1.0010404	1.0012067
18	1.001184	1.0012132	1.0013900
19	1.001370	1.0013965	1.0015800
20	1.001567	1.0015940	1.0017900
21	1.001776	1.0017997	1.0020000
22	1.001995	1.0020108	1.0022200
23	1.002225	1.0022310	1.0024400
24	1.002465	1.0024648	1.0027100
25	1.002715	1.0027075	1.0029300
26		1.0029588	1.0032100
27		1.0032211	1.0034500
28		1.0034944	1.0037400
29		1.0037758	1.0040300
30	1.004064	1.0040710	1.0043300
35	1.005697	1.0056770	1.0059300
40	1.007531	1.0075120	1.0077300
45	1.009541	1.0095625	1.0098500
50	1.011766	1.0118150	1.0120500
55	1.014100	1.0143596	1.0144500
60	1.016590	1.0171180	1.0169800
65	1.019302	1.0199465	1.0196700
70	1.022246	1.0229376	1.0225500
75	1.025440	1.0260782	1.0256200
80	1.028581	1.0293600	1.0288500
85	1.031894	1.0327692	1.0322500
90	1.035397	1.0362943	1.0356600
95	1.039094	1.0399247	1.0392500
+100	1.042986	1.0436490	1.0431500

TABLE

OF THE TENSION OF THE VAPOR OF ABSOLUTE ALCOHOL.
Calculated from Regnault's data by Bunsen.

Centi-grade	Tenths of Degrees.									
	0.0	0.1	0.2	0.3	0.4	0.5	0.6	0.7	0.8	0.9
0	mm. 12.73	12.82	12.91	13.01	13.10	13.19	13.28	13.37	13.46	13.56
1	13.65	13.74	13.84	13.93	14.03	14.12	14.22	14.31	14.41	14.50
2	14.60	14.70	14.79	14.89	14.99	15.09	15.19	15.29	15.39	15.49
3	15.59	15.69	15.79	15.90	16.00	16.10	16.21	16.31	16.41	16.52
4	16.62	16.73	16.84	16.95	17.05	17.16	17.27	17.38	17.48	17.59
5	17.70	17.82	17.93	18.04	18.16	18.27	18.38	18.50	18.61	18.73
6	18.84	18.96	19.08	19.20	19.32	19.44	19.56	19.68	19.80	19.92
7	20.04	20.17	20.30	20.43	20.55	20.68	20.81	20.93	21.06	21.19
8	21.31	21.45	21.58	21.72	21.85	21.99	22.12	22.25	22.39	22.52
9	22.66	22.80	22.94	23.08	23.23	23.37	23.51	23.65	23.79	23.94
10	24.08	24.23	24.38	24.53	24.68	24.83	24.99	25.14	25.29	25.44
11	25.59	25.75	25.91	26.07	26.23	26.39	26.55	26.71	26.87	27.03
12	27.19	27.36	27.53	27.70	27.87	28.04	28.21	28.38	28.55	28.72
13	28.89	29.07	29.25	29.43	29.61	29.79	29.97	30.15	30.23	30.51
14	30.69	30.88	31.07	31.26	31.45	31.64	31.84	32.03	32.22	32.41
15	32.60	32.80	33.01	33.21	33.41	33.61	33.82	34.02	34.22	34.42
16	34.62	34.84	35.05	35.27	35.48	35.70	35.91	36.13	36.34	36.56
17	36.77	37.00	37.23	37.45	37.68	37.91	38.14	38.36	38.59	38.82
18	39.05	39.29	39.53	39.77	40.01	40.25	40.49	40.73	40.97	41.21
19	41.45	41.71	41.96	42.22	42.47	42.73	42.98	43.24	43.49	43.75
20	44.00	44.27	44.54	44.81	45.08	45.35	45.61	45.88	46.15	46.42
21	46.69	46.98	47.26	47.55	47.83	48.12	48.40	48.69	48.97	49.26
22	49.54	49.84	50.14	50.44	50.74	51.04	51.34	51.64	51.94	52.24
23	52.54	52.86	53.17	53.49	53.81	54.12	54.44	54.75	55.07	55.38
24	55.70	56.04	56.37	56.70	57.03	57.37	57.70	58.03	58.36	58.70
25	59.03	59.38	59.73	60.08	60.43	60.78	61.13	61.48	61.83	62.18
26	62.53	62.90	63.27	63.64	64.01	64.37	64.74	65.11	65.48	65.85
27	66.22	66.60	66.99	67.38	67.77	68.15	68.54	68.93	69.31	69.70
28	70.02	70.49	70.89	71.29	71.69	72.09	72.49	72.89	73.29	73.69
29	74.09	74.53	74.96	75.39	75.82	76.25	76.68	77.12	77.55	77.98
30	78.40									

TENSION OF THE VAPOR OF ALCOHOL.

REGNAULT. — Mémoires de l'Académie de France, Tome XXVI. p. 374.

In this and the following Tables of Tensions of Vapor the Temperature is in degrees of the Air Thermometer.

Temperature.	Tension.	Temperature.	Tension.	Temperature.	Tension.	Temperature.	Tension.	Temperature.	Tension.
o	mm.	o	mm.	o	mm.	o	mm.		mm.
—20	3.34	20	44.46	60	350.21	95	1425.13	135	4964.22
—15	5.10	25	59.37	65	436.90	100	1697.55	140	5674.59
—10	6.47	30	78.52	70	541.15	105	2010.38	145	6458.10
— 5	9.09	35	102.91	75	665.54	110	2367.64	150	7318.40
0	12.70	40	133.69	78.26	760.00	115	2773.40	155	8259.19
+ 5	17.62	45	172.18	80	812.91	120	3231.73		
10	24.23	50	219.90	85	986.40	125	3746.88		
15	32.98	55	278.59	90	1189.30	130	4323.00		

TENSION OF THE VAPOR OF ETHER.

REGNAULT. — Mémoires de l'Académie de France, Tome XXVI. p. 392.

Temperature	Tension	Temperature.	Tension.	Temperature.	Tension.	Temperature.	Tension.	Temperature.	Tension.
o	mm.	o	mm.	o	mm.	o	mm.	o	mm.
—20	68.90	10	286.83	35	761.20	65	1998.87	95	4401.81
—15	89.31	15	353.62	40	907.04	70	2304.90	100	4953.30
—10	114.72	20	432.78	45	1074.15	75	2645.41	105	5556.23
— 5	146.06	25	525.93	50	1264.83	80	3022.79	110	6214.63
0	184.39	30	634.80	55	1481.06	85	3439.53	115	6933.26
+ 5	230.89	34.97	760.00	60	1725.01	90	3898.26	120	7719.20

TENSION OF THE VAPOR OF BISULPHIDE OF CARBON.

REGNAULT. — Mémoires de l'Académie de France, Tome XXVI. p. 402.

Temperature.	Tension.	Temperature.	Tension.	Temperature.	Tension.	Temperature.	Tension.	Temperature.	Tension.
o	mm.	o	mm.	o	mm.	95	mm.	o	mm.
—20	47.30	20	298.03	55	1001.57	95	2966.34	135	6925.90
—15	61.64	25	361.13	60	1164.51	100	3325.15	140	7603.96
—10	79.44	30	434.62	65	1347.52	105	3727.19	145	8326.92
— 5	101.29	35	519.66	70	1552.09	110	4164.06	150	9095.94
0	127.91	40	617.53	75	1779.88	115	4637.41		
+ 5	160.01	45	729.53	80	2032.53	120	5148.79		
10	198.46	46.20	760.00	85	2311.70	125	5699.69		
15	244.13	50	857.07	90	2619.08	130	6291.60		

TENSION OF THE VAPOR OF CHLOROFORM.

REGNAULT. — Mémoires de l'Académie de France, Tome XXVI. p. 415.

Temperature.	Tension.	Temperature.	Tension.	Temperature.	Tension.	Temperature.	Tension.	Temperature.	Tension.
°	mm.	°	mm.	°	mm.	°	mm.	°	mm.
+20	160.47	55	637.71	85	1624.10	120	3925.74	155	7985.35
25	200.18	60	755.44	90	1865.22	125	4386.60	160	8734.20
30	247.51	60.16	760.00	95	2132.85	130	4885.10	165	9527.82
35	303.49	65	889.72	100	2428.54	135	5422.53		
40	369.26	70	1042.11	105	2754.03	140	6000.16		
45	446.01	75	1214.20	110	3110.99	145	6619.20		
50	535.05	80	1407.64	115	3501.03	150	7280.62		

TENSION OF THE VAPOR OF BENZINE.

REGNAULT. — Mémoires de l'Académie de France, Tome XXVI. p. 428.

Temperature.	Tension.	Temperature.	Tension.	Temperature.	Tension.	Temperature.	Tension.	Temperature.	Tension.
°	mm.	°	mm.	°	mm.	°	mm.	°	mm.
—20	5.79	20	75.65	60	390.10	95	1167.46	135	3158.51
—15	8.82	25	95.91	65	463.43	100	1340.05	140	3520.73
—10	12.92	30	120.24	70	547.42	105	1531.83	145	3912.11
— 5	18.33	35	149.26	75	643.18	110	1744.12	150	4333.71
0	25.31	40	183.62	80	751.86	115	1978.22		
+ 5	34.17	45	224.06	80.36	760.00	120	2235.44		
10	45.25	50	271.37	85	874.63	125	2517.06		
15	58.93	55	326.41	90	1012.75	130	2824.35		

TENSION OF THE VAPOR OF CHLORIDE OF CARBON, (C_2Cl_6).

REGNAULT. — Mémoires de l'Académie de France, Tome XXVI. p. 439.

Temperature.	Tension.	Temperature.	Tension.	Temperature.	Tension.	Temperature.	Tension.	Temperature.	Tension.
°		°	mm.	°	mm.	°	mm.	°	mm.
—20	9.80	25	114.30	70	621.15	110	1887.44	155	5010.21
—15	13.55	30	142.27	75	725.66	115	2129.15	160	5513.14
—10	18.47	35	175.55	76.5	760.00	120	2393.67	165	6053.83
— 5	24.83	40	214.81	80	843.29	125	2682.41	170	6634.37
0	32.95	45	260.82	85	975.12	130	2996.88	175	7256.87
+ 5	43.19	50	314.38	90	1122.26	135	3338.56	180	7923.55
10	55.97	55	376.29	95	1286.86	140	3709.04	185	8636.78
15	71.73	60	447.43	100	1467.09	145	4109.99	190	9399.02
20	90.99	65	528.74	105	1667.19	150	4543.13		

TENSION OF THE VAPOR OF CHLORIDE OF ETHYL.

REGNAULT. — Mémoires de l'Académie de France, Tome XXVI. p. 446.

Temperature.	Tension.	Temperature.	Tension.	Temperature.	Tension.	Temperature.	Tension.	Temperature.	Tension.
°	mm.	°	mm	°	mm.	°	mm.	°	mm.
—30	110.24	0	465.18	25	1184.17	55	2668.43	85	6301.61
—25	145.01	+ 5	569.32	30	1398.99	60	3400.54	90	7047.51
—20	187.85	10	691.11	35	1643.24	65	3878.52	95	7853.92
—15	239.60	12.5	760.00	40	1619.58	70	4405.03	100	8722.76
—10	302.09	15	832.56	45	2230.71	75	4982.72		
— 5	376.72	20	996.23	50	2579.40	80	5614.11		

TENSION OF THE VAPOR OF BROMIDE OF ETHYL.

REGNAULT. — Mémoires de l'Académie de France, Tome XXVI. p. 453.

Temperature.	Tension.	Temperature.	Tension.	Temperature.	Tension.	Temperature.	Tension.	Temperature.	Tension.
°	mm.	°	mm.	°	mm.	°	mm.	°	mm.
—30	32.18	10	257.40	45	947.28	85	3000.63	125	7362.25
—25	44.06	15	316.92	50	1112.79	90	3398.95	130	8116.49
—20	59.16	20	387.03	55	1300.35	95	3835.53	135	8921.92
—15	78.09	25	469.07	60	1511.92	100	4312.32	140	9779.56
—10	101.54	30	564.51	65	1749.47	105	4831.22		
— 5	130.58	35	674.92	70	2015.06	110	5394.01		
0	665.57	38.37	760.00	75	2310.73	115	6002.41		
+ 5	207.21	40	801.92	80	2638.57	120	6658.00		

TENSION OF THE VAPOR OF IODIDE OF ETHYL.

REGNAULT. — Mémoires de l'Académie de France, Tome XXVI. p. 456.

Temperature.	Tension.	Temperature.	Tension.	Temperature.	Tension.	Temperature.	Tension.	Temperature.	Tension.
°	mm.	°	mm.	°	mm.	°	mm.	°	mm.
0	41.95	15	87.64	30	169.07	45	303.77	60	512.25
15	54.14	20	110.02	35	207.09	50	364.00		
10	69.20	25	136.95	40	251.73	55	433.21		

TENSION OF THE VAPOR OF WOOD SPIRIT.

REGNAULT. — Mémoires de l'Académie de France, Tome XXVI. p. 460.

Temperature.	Tension.	Temperature.	Tension.	Temperature.	Tension.	Temperature.	Tension.	Temperature.	Tension.
°	mm.	°	mm.	°	mm.	°	mm.	°	mm.
—30	2.67	10	50.13	50	381.68	85	1470.92	125	4980.55
—25	4.14	15	67.11	55	472.20	90	1741.67	130	5691.30
—20	6.27	20	88.67	60	579.93	95	2051.71	135	6479.32
—15	9.29	25	115.99	65	707.33	100	2405.15	140	7337.10
—10	13.47	30	149.99	66.78	760.00	105	2806.27	145	8308.87
— 5	19.17	35	192.01	70	857.10	110	3259.60	150	9361.35
0	26.82	40	243.51	75	1032.14	115	3769.80		
+ 5	36.89	45	306.13	80	1238.47	120	4341.77		

TENSION OF THE VAPOR OF ACETONE.

REGNAULT. — Mémoires de l'Académie de France, Tome XXVI. p. 472.

Temperature.	Tension.	Temperature.	Tension.	Temperature.	Tension.	Temperature.	Tension.	Temperature.	Tension.
°	mm.	°	mm.	°	mm.	°	mm.	°	mm.
+20	179.63	50	602.86	70	1189.38	95	2452.81	120	4546.86
25	226.27	55	725.95	75	1387.62	100	2797.27	125	5086.25
30	281.00	56.3	760.00	80	1611.05	105	3177.00	130	5669.72
35	345.15	60	860.48	85	1861.81	110	3593.96	135	6298.68
40	420.15	65	1014.32	90	2141.66	115	4050.02	140	6974.43
45	507.52								

TENSION OF THE VAPOR OF BROMIDE OF ETHYLENE.

REGNAULT. — Mémoires de l'Académie de France, Tome XXVI. p. 468.

Temperature.	Tension.	Temperature.	Tension.	Temperature.	Tension.	Temperature.	Tension.	Temperature.	Tension.
°	mm.	°	mm.	°	mm.	°	mm.	°	mm.
—25	1.55	30	17.20	85	172.92	135	833.26	190	3020.83
—20	1.73	35	21.80	90	206.58	140	953.00	195	3833.45
—15	2.03	40	27.49	95	245.51	145	1085.89	200	3668.36
—10	2.48	45	34.47	100	290.43	150	1232.83	205	4026.25
— 5	3.09	50	42.99	105	342.11	155	1394.73	210	4407.52
0	3.92	55	53.31	110	401.08	160	1572.49	215	4812.84
+ 5	5.01	60	65.75	115	468.13	165	1766.99	220	5242.61
10	6.42	65	80.64	120	544.06	170	1979.14	225	5597.17
15	8.25	70	98.36	125	629.66	175	2209.77	230	6176.87
20	10.57	75	119.34	130	725.77	180	2459.73	235	6681.92
25	13.51	80	144.02	131.6	760.00	185	2729.84	240	7212.51

TENSION OF THE VAPOR OF CHLORIDE OF SILICON (SiCl₄).

REGNAULT. — Mémoires de l'Académie de France, Tome XXVI. p. 476.

Temperature.	Tension.	Temperature.	Tension.	Temperature.	Tension.	Temperature.	Tension.	Temperature.	Tension.
o	mm.	o	mm.	o	mm.	o	mm.	o	mm.
—25	19.66	— 5	60.52	15	157.74	35	356.83	55	715.44
—20	26.49	0	78.62	20	195.86	40	429.08	56.81	760.00
—15	35.28	+ 5	99.59	25	241.15	45	512.32	60	837.23
—10	46.46	10	125.90	30	294.49	50	607.46	65	973.74

TENSION OF THE VAPOR OF TERCHLORIDE OF PHOSPHORUS.

REGNAULT. — Mémoires de l'Académie de France, Tome XXVI p. 478.

Temperature.	Tension.	Temperature.	Tension.	Temperature.	Tension.	Temperature.	Tension.	Temperature.	Tension.
o	mm.	o	mm.	o	mm.	o	mm.	o	mm.
0	37.98	20	100.55	40	233.78	55	408.46	70	674.23
+ 5	49.09	25	125.59	45	283.46	60	485.63	73.80	760.00
10	62.88	30	155.65	50	341.39	65	573.86	75	787.61
15	79.85	35	191.45						

TENSION OF THE VAPOR OF TERCHLORIDE OF BORON.

REGNAULT. — Mémoires de l'Académie de France, Tome XXVI. p. 480.

Temperature.	Tension.	Temperature.	Tension.	Temperature.	Tension.	Temperature.	Tension.	Temperature.	Tension.
o	mm.	o	mm.	o	mm.	o	mm.	o	mm.
—30	98.25	— 5	310.30	18.23	760.00	40	1535.25	65	3010.24
—25	125.68	0	381.32	20	807.50	45	1775.69	70	3392.12
—20	159.46	+ 5	465.03	25	957.29	50	2042.25	75	3804.79
—15	200.69	10	562.94	30	1127.50	55	2336.17	80	4248.28
—10	250.54	15	676.57	35	1319.66	60	2658.52	85	4720.11

TENSION OF THE VAPOR OF CHLORIDE OF CYANOGEN.

REGNAULT. — Mémoires de l'Académie de France, Tome XXVI. p. 484.

Temperature.	Tension.	Temperature.	Tension.	Temperature.	Tension.	Temperature.	Tension.	Temperature.	Tension.
°	mm.	°	mm.	°	mm.	°	mm.	°	mm.
—30	68.30	— 5	350.20	15	830.30	40	1987.96	65	4232.24
—25	103.38	0	444.11	20	1001.87	45	2329.77	70	4873.19
—20	148.21	+ 5	553.99	25	1199.76	50	2719.29	75	5594.58
—15	203.58	10	681.92	30	1427.43	55	3162.11		
—10	270.51	12.66	760.00	35	1688.74	60	3664.24		

TENSION OF THE VAPOR OF OIL OF TURPENTINE.

REGNAULT. — Mémoires de l'Académie de France, Tome XXVI. p. 501.

Temperature.	Tension.	Temperature.	Tension.	Temperature.	Tension.	Temperature.	Tension.	Temperature.	Tension.
°	mm.	°	mm.	°	mm.	°	mm.	°	mm.
0	2.07	60	26.46	120	257.21	160	775.09	190	1473.24
10	2.94	70	40.64	130	348.98	165	871.27	195	1618.26
20	4.45	80	61.30	140	464.02	170	975.42	200	1771.47
30	6.87	90	90.61	150	605.20	175	1090.11		
40	10.80	100	131.11	155	686.37	180	1207.92		
50	16.98	110	185.62	159.2	760.00	185	1336.45		

TENSION OF THE VAPOR OF SULPHUR.

REGNAULT. — Mémoires de l'Académie de France, Tome XXVI. p. 530.

Temperature.	Tension.	Temperature.	Tension.	Temperature.	Tension.	Temperature.	Tension.	Temperature.	Tension.
°	mm.	°	mm.	°	mm.	°	mm.	°	mm.
390	272.31	430	560.98	470	1063.17	510	1871.57	550	3086.51
400	328.98	440	663.11	480	1232.70	520	2133.30	560	3465.33
410	395.20	450	779.89	490	1422.88	530	2421.97	570	3877.08
420	472.11	460	912.74	500	1635.32	540	2739.21		

TENSION OF THE VAPOR OF SULPHUROUS ACID.

REGNAULT. — Mémoires de l'Académie de France, Tome XXVI. p. 590.

Temperature.	Tension.	Temperature.	Tension.	Temperature.	Tension.	Temperature.	Tension.	Temperature.	Tension.
°	mm.	°	mm.	°	mm.	°	mm.	°	mm.
—30	287.47	—10	762.49	10	1799.55	30	3431.80	50	6220.01
—25	373.79	— 5	946.90	15	2064.90	35	4014.78	55	7125.02
—20	479.46	0	1165.06	20	2462.05	40	4670.23	60	8123.80
—15	607.90	+ 5	1421.14	25	2915.97	45	5403.52	65	9221.40
10.08	760.00								

TENSION OF THE VAPOR OF OXIDE OF METHYL.

REGNAULT. — Mémoires de l'Académie de France, Tome XXVI. p. 593.

Temperature.	Tension.	Temperature.	Tension.	Temperature.	Tension.	Temperature.	Tension.	Temperature.	Tension.
°	mm.	°	mm.	°	mm.	°	mm.	°	mm.
—30	576.54	—20	882.00	—5	1572.51	10	2628.97	25.	4151.00
—25	716.08	—15	1077.67	0	1879.02	15	3079.80	30	4777.99
23.65	760.00	—10	1306.63	+5	2229.93	20	3586.01		

TENSION OF THE VAPOR OF CHLORIDE OF METHYL.

REGNAULT. — Mémoires de l'Académie de France, Tome XXVI. p. 595.

Temperature.	Tension.	Temperature.	Tension.	Temperature.	Tension.	Temperature.	Tension.	Temperature.	Tension.
°	mm.	°	mm.	°	mm.	°	mm.	°	mm.
—30	578.99	—20	883.25	— 5	1578.70	10	2663.81	25	4267.40
—25	717.76	—15	1079.20	0	1891.00	15	3134.10	30	4940.46
23.73	760.00	—10	1309.61	+ 5	2251.08	20	3666.95	35	5691.08

TENSION OF THE VAPOR OF AMMONIA.

REGNAULT. — Mémoires de l'Académie de France, Tome XXVI. p. 607.

Temperature.	Tension.	Temperature.	Tension.	Temperature.	Tension.	Temperature.	Tension.	Temperature.	Tension.
°	mm.	°	mm	°	mm.	°	mm.	°	mm.
-38.5	760.00	— 5	2624.22	25	7477.00	55	17219.78	85	34330.87
—30	866.09	0	3183.34	30	8700.97	60	19482.10	90	38109.22
—25	1104.28	+ 5	3830.34	35	10070.18	65	21965.13	95	42195.71
—20	1392.13	10	4574.03	40	11595.30	70	24675.55	100	46608.24
—15	1736.48	15	5123.40	45	13287.31	75	27629.98		
—10	2144.62	20	6387.78	50	15158.33	80	30843.09		

TENSION OF THE VAPOR OF SULPHURETTED HYDROGEN.

REGNAULT. — Mémoires de l'Académie de France, Tome XXVI. p. 616.

Temperature.	Tension.	Temperature.	Tension.	Temperature.	Tension.	Temperature.	Tension.	Temperature.	Tension.
°	mm.	°	mm.	°	mm.	°	mm.	°	mm.
−61.8	760.00	− 5	7066.03	15	12447.94	35	20224.31	55	30690.69
−25	3749.33	0	8206.29	20	14151.51	40	22582.46	60	33740.16
−20	4438.45	+ 5	9490.80	25	16012.39	45	24954.26	65	36961.55
−15	5196.52	10	10896.32	30	18035.35	50	27814.77	70	40353.25
−10	6084.57								

TENSION OF THE VAPOR OF CARBONIC ACID.

REGNAULT. — Mémoires de l'Académie de France, Tome XXVI. p. 625.

Temperature.	Tension.	Temperature.	Tension.	Temperature.	Tension.	Temperature.	Tension.	Temperature.	Tension.
°	mm.	°	mm.	°	mm.	°	mm.	°	mm.
78.2	760.00	−10	20340.20	+ 5	30753.80	20	44716.58	35	62447.30
−25	13007.02	− 5	23441.34	10	34998.65	25	50207.32	40	69184.45
−20	15142.44	0	26906.60	15	39646.86	30	56119.05	45	73314.60
−15	17582.48								

TENSION OF THE VAPOR OF PROTOXIDE OF NITROGEN.

REGNAULT. — Mémoires de l'Académie de France, Tome XXVI. p. 631.

Temperature.	Tension.	Temperature.	Tension.	Temperature.	Tension.	Temperature.	Tension.	Temperature.	Tension.
°	mm.	°	mm.	°	mm.	°	mm.	°	mm.
87.9	760.00	−15	19684.33	0	27420.97	15	37831.66	30	51708.55
−25	15694.88	−10	22008.05	+ 5	30558.64	20	42027.88	35	57268.08
−20	17586.58	− 5	24579.20	10	34019.09	25	46641.40	40	63359.78

NOTE. — Tables for tension of vapor of Water and Mercury will be found in the next section.

TABLE

OF UNITS OF HEAT EVOLVED BY THE COMBINATION OF EQUIVA-
LENTS, IN GRAMMES, OF SOME OF THE ELEMENTS.

*The quantity of Heat necessary to raise 1 gramme of Water at 0° C. one
degree, being unity. Hydrogen = 1. Those marked F. S. are by Favre
and Silbermann, A. by Andrews. Those marked * were calculated, the
others observed.*

MILLER. — Chem. Physics, p. 309.

Elements.	Observers.	Oxygen.	Chlorine.	Bromine.	Iodine.	Sulphur.
Hydrogen	F. S.	34462	23783	*9322	*3606	*2741
Carbon	F. S.	24240				
Sulphur	F. S.	17760				
Phosphorus	A.	36072				
Potassium	A.		104476			
"	F. S.		*100960	*90188	*77268	*45638
Sodium	F. S.		94847			
Zinc	A.	42282	50658	40640	26617	
"	F. S.	*42451	*50296			*20940
Iron	A.	33072	32695	23833	8046	
"	•F. S.	*37828	*49651			*17753
Tin	A.	33519	*31722			
Arsenic	A.		24992			
Antimony	Dulong	47000	A. 30401			
Copper	Dulong	19152	30404			
"	F. S.	*21885	*29524			*9133
Lead	F. S.	*27675	*44730	*32802	*23208	*9556
Silver	F. S.	*6113	*34800	*25618	*18651	*5524

HEAT EVOLVED DURING COMBUSTION IN OXYGEN.

Heat unit = 1 gramme of water at 0° C. raised 1°.

MILLER. — Chem. Physics, p. 299.

Substance burned.	Heat units evolved by combustion of 1 gramme of substance.	Heat units evolved by combination of 1 gramme of Oxygen.	Heat units evolved by 8 grammes (1 eqt.) of Oxygen.	Compound formed.	Observers.
Hydrogen	34462	4307	34462	HO	F. & S.*
"	34743	4343	34743	"	Dulong
Carbon	8080	3030	24240	CO_2	F. & S.
"	7912	6967	23736	"	Despretz
Sulphur	2220	2220	17760	SO_2	F. & S.
"	2307	2307	18456	"	Andrews
"	2601	2601	20808	"	Dulong
Phosphorus	5747	4509	36072	PO_5	Andrews
"	5669	4394	35148	"	Abria
Zinc	1301	5285	42282	ZnO	Andrews
"	1298	5273	42185	"	Dulong
Iron	1576	4134	33072	Fe_3O_4	Andrews
"	1702	4340	34720	"	Dulong
Cobalt	1080	3995	31960	?	"
Nickel	1006	3723	29784	NiO	"
Tin	1233	4545	36360	SnO_2	"
"	1144	4230	33519	"	Andrews
Antimony	961	5875	47000	SbO_4	Dulong
Copper	602	2394	19152	CuO	Andrews
"	632	2512	20096	"	Dulong
Carbonic Oxide	2634	4609	36876	CO_2	"
"　　　　"	2431	4258	34034	"	Andrews
"　　　　"	2403	4205	33642	"	F. & S.
Protoxide of Tin	534	4473	35784	SnO_2	Dulong
"　　　　"	521	4349	34792	"	Andrews
Suboxide of Copper	256	2288	18304	CuO	"
"　　　　"	244	2185	17480	"	Dulong
Cyanogen	5195	4221	33768	"	"
Marsh Gas	13063	3266	26128		F. & S.
"　　　　"	13185	3296	26368		Dulong
Olefiant Gas	11942	3483	27864		Andrews
"　　　　"	11858	3458	27664		F. & S.
"　　　　"	12030	3514	28112		Dulong
Alcohol	6909	3311	26488		"
"	6850	3282	26256		Andrews
"	7183	3442	27536		F. & S.
Ether	9027	3480	27840		"
Olive Oil	9862				Dulong
Oil of Turpentine	10852	3294	26352		F. & S.
Bisulphide of Carbon	3401	2692	21536		"

* F. & S., Favre and Silbermann.

QUANTITIES OF HEAT DISENGAGED BY THE ACTION OF BROMINE AND IODINE.

MILLER. — Chem. Physics, p. 300.

Elements.	Heat units evolved by combination of 1 gramme of the substance.	Heat units evolved by combination of 1 gramme of Iodine or Bromine.	Heat units evolved by combination of 1 eqt. of Bromine or Iodine.	Compound produced.	Observer.
Bromine.					
Zinc	1269	508	40640	ZnBr	Andrews
Iron	1277	298	23833	Fe_2Br_3	"
Iodine.					
Zinc	819	209	26617	ZnI	Andrews
Iron	463	63	8046	Fe_2I_3	"

QUANTITIES OF HEAT DISENGAGED BY THE ACTION OF CHLORINE.

MILLER. — Chem. Physics, p. 300.

Elements.	Heat units evolved by combination of 1 gramme of the Element.	Heat units evolved by combination of 1 gramme of Chlorine.	Heat units evolved by combination of 1 eqt. of Chlorine.	Compound produced.	Observer.
Hydrogen	24087	678	24087	HCl	Abria
"	23783	670	23783	"	F. & S.
Phosphorus	3422 ?	607	21548	PCl_5 ?	Andrews
Potassium	2655	2943	104476	KCl	"
Zinc	1529	1427	50658	ZnCl	"
Iron	1745	921	32695	Fe_2Cl_3	"
Tin	1079	897	31722	$SnCl_2$	"
Antimony	707	860	30401	$SbCl_3$	"
Arsenic	994	704	24992	$AsCl_3$	"
Copper	961	859	30494	CuCl	"
Mercury	?	822	29181	?	"

HEAT UNITS EVOLVED BY COMBINATION OF ONE EQUIVALENT OF THE UNDERMENTIONED BASES WITH ONE EQUIVALENT OF CERTAIN ACIDS.

MILLER. — Chem. Physics, pp. 310, 311.

Aci	Observers	Potassium.	Sodium.	Ammonium.	Magnesium.	Zinc.	Manganese.	Nickel.	Cobalt.	Iron (ferrous).	Copper.
Sulphuric	F. & S.	16083	15810	14690	14440	10455	12075	11932	11780	10872	7720
"	A.	15900	16200	13900	18500	11800					
Nitric	F. & S.	15510	15283	13676	12840	8323	10850	10450	9956	9648	6400
"	A.	14600	14000	12200	17700	10300					
Phosphoric	F. & S.	17766									
"	A.	14200	14000	12200							
Arsenic	A.	14100	13900	12200							
Hydrochloric	A.	14300	14700								
"	F. & S.	15656	15128	13536	17700	10600	11285	10412	10374	9828	6416
Hydrobromic	F. & S.	15510	15159		13220	8307					
Hydriodic	F. & S.	15698	15097								
"	A.	14200	14100	14700							
Oxalic	A.	14400	14600	12400	11900	9800					
"	F. & S.	14156	13752								
Acetic	F. & S.	13973	13600	12649	12270	7720	9982	9245	9272	8590	5264
Formic	F. & S.	13900	13308								
"	A.										
Tartaric	A.	13200	12900	11200							
"	F. & S.	13425	12651								
Citric	A.	13200	12900	11000							
"	F. & S.	13658	13178								
Succinic	A.	13400	12900	11200							

F. & S., Favre and Silbermann ; A., Andrews.

TABLES

FOR GAS ANALYSIS.

FORMULAS

FOR THE WEIGHTS, VOLUMES, AND DENSITIES OF GASES.

Volume of a gas at 0° C. and 760 mm. when its volume at a given temperature and pressure is known.

$$V^0 = V' \frac{1}{1 + 0.00367\,t} \times \frac{h}{760}.$$

V^0 = volume at 0° C. 760 mm.,
V' = the observed volume,
$h =$ " " pressure in mm. of mercury reduced to 0° C.

Density at 0° C. and 760 mm. when the density at a given temperature and pressure is known.

$$D^0 = D' \, (1 + 0.00367\,t) \, \frac{760}{h}.$$

D^0 = density at 0° C. and 760 mm.,
$D^\cdot$ = observed density.

To determine the density of a gas or vapor by Gay-Lussac's method.

$$D = \frac{w}{0.001293 \, V \frac{1}{1 + 0.00367\,t} \times \frac{h}{760}}.$$

w = the weight of the vapor,
V = the observed volume,
$h =$ " " pressure in mm. of mercury at 0° C.,
t = temperature of the vapor,
0.001293 = weight of 1 cubic centimetre of air at 0° C. and 760 mm.

To determine the density of a vapor by Dumas's method.

$$D = \frac{W' - (W - w)}{0.001293 \, V \frac{1}{1 + 0.00367\,v} \times \frac{h}{760}}.$$

When there is an air bubble

$$D = \frac{W' - (W - w) - w'}{0.001293 \, (V - v') \frac{1}{1 + 0.00367\,t} \times \frac{h}{760}}.$$

In which formulas

$$w = 0.001293 \ V \frac{1}{1 + 0.00367 \, t} \times \frac{h'}{760}.$$

$$w' = 0.001293 \ (V - V') \frac{1}{1 + 0.00367 \, t''} \times \frac{h''}{760}.$$

$$v' = (V - V') \frac{1 + 0.00367 \, t'}{1 + 0.00367 \, t''} \times \frac{h''}{h}.$$

w = weight of air in balloon when first weighed,
w' = weight of air bubble,
v' = volume of air bubble,
W = weight of balloon filled with air,
t = temperature at moment of weighing,
h' = the height of the barometer,
t' = temperature at moment of sealing,
h = height of barometer at moment of sealing,
W' = weight of balloon filled with vapor,
V = capacity of balloon in cubic centimetres,
V' = the volume of water or mercury which enters the balloon.
$(V - V')$ = volume of air bubble,
t'' = the temperature during this determination,
h'' = pressure on enclosed bubble of air.

In almost all cases the factors into which the pressures enter may be neglected, since they may be regarded as constant during the experiment. If the expansion of the glass is allowed for, the formula becomes

$$D = \frac{W' - (W - w) - w'}{0.0012932 \ (V a - v) \frac{1}{1 + 0.00367 \, t'} \times \frac{h}{760}}.$$

In which, $a = 1 + k \ (t' - t'')$, k being the coefficient of expansion of the glass.

The above formula may be written

$$D = \frac{W' - W + (A \times V \times h') - (A \times (V - V') \times h'')}{\left[V a - (V - V'') \ B \ (t' - t'') \frac{h''}{h} \right] A' \times h}$$

$$A = \frac{0.0012932}{760 \ (1 + 0.00367 \, t)}. \qquad\qquad B = 1 + \frac{0.00367}{1 + 0.00367 \, t''}.$$

$$A' = \frac{0.0012932}{760\,(1 + 0.00367\,t')}. \qquad a = 1 + k\,(t' - t'').$$

Which values will be found in the tables. Generally $\dfrac{h''}{h'}$ and a may be neglected, as they are very small.

Deville and Troost use the following constants in determining vapor densities.

				cc.
Vapor of Boiling	Mercury			350
"	"	"	Sulphur	440
"	"	"	Cadmium	860
"	"	"	Zinc	1040

The foregoing formulas by the use of logarithms become

For volume of a gas

$$\text{Log } V^0 = \text{Log } V' + \text{Log } \frac{1}{1 + 0.00367\,t} + \text{Log } \frac{h}{760}.$$

For Logs of $\dfrac{1}{1 + 0.00367\,t}$ and $\text{Log } \dfrac{h}{760}$ see Tables.

For weight

$$\text{Log weight} = \text{Log } w + \text{Log } V' + \text{Log } \frac{1}{1 + 0.00367\,t} + \text{Log } \frac{h}{760}.$$

$w =$ weight of one cubic centimetre of the gas. (Page 93.)

For the density of a gas at 0° C. and 760 mm.

$$\text{Log } D^0 = \text{Log } D' + \text{Log } (1 + 0.00367\,t) + \text{Log } 760 - \text{Log } h,$$

$$\text{Log } 760 = 2.88081.$$

For the density of a vapor by Gay-Lussac's method

$$\text{Log } D = \text{Log } w - \left[\text{Log } 0.001293 + \text{Log } V' + \text{Log } \left(\frac{1}{1 + 0.00367\,t} \right) \right.$$
$$\left. + \text{Log } \left(\frac{h}{760} \right) \right].$$

Log .001293 = 7.11166.

For the density of a vapor by Dumas's method

$$\text{Log } D = \text{Log } [W' - (W - w)] - \left[\text{Log } 0.001293 + \text{Log } V \right.$$
$$\left. + \text{Log } \left(\frac{1}{1 + 0.00367\, t'} \right) + \text{Log } \left(\frac{h}{760} \right) \right].$$

When there is an air bubble

$$\text{Log } D = \text{Log } [W' - (W - w) - w'] - \left[\text{Log } 0.001293 + \text{Log } (V - v') \right.$$
$$\left. + \text{Log } \left(\frac{1}{1 + 0.00367\, t'} \right) + \text{Log } \left(\frac{h}{760} \right) \right].$$

$$\text{Log } w = \text{Log } 0.001293 + \text{Log } V + \text{Log } \left(\frac{1}{1 + 0.00367\, t} \right) + \text{Log } \left(\frac{h'}{760} \right).$$

$$\text{Log } w' = \text{Log } 0.001293 + \text{Log } (V - V') + \text{Log } \left(\frac{1}{1 + 0.00367\, t''} \right)$$
$$+ \text{Log } \left(\frac{h''}{760} \right).$$

$$\text{Log } v' = \text{Log } (V - V') + \text{Log } (1 + 0.00367\, t') + \text{Log } h''$$
$$+ \text{Log } \left(\frac{1}{1 + 0.00367\, t''} \right) + \text{Log } \left(\frac{1}{h} \right).$$

In most cases it will be sufficient to take the value of h to the nearest millimetre.

TABLE

OF PHYSICAL PROPERTIES OF SOME OF THE MOST COMMON GASES.

REGNAULT.—Mémoires de l'Académie Royale de France, Tome XXI.

Name.	Density.		Coefficients of Expansion for 100° C.		Weight of 1 cc. 0° C. 760 m.	Log. of the weight of 1 cc. of the gas.
	Air = 1 at 0° C. 760 m.	Water = 1 at 4° 0. 760 m.	Constant Volume.	Constant Pressure.		
Air	1.00000	0.001293187	0.3665	0.3670	0.001293187	7.11166
Hydrogen	0.06926	0.000089578	0.3667	0.3661	0.000089578	5.95220
Nitrogen	0.97200	0.001256167	0.3668	0.3670	0.001256167	7.09905
Oxygen	1.10570	0.001429802			0.001529802	7.18464
Carbonic oxide			0.3667	0.3669		
" acid	1.52901	0.00197414	0.3688	0.3710	0.001977414	7.29611
Protoxide of Nitrogen			0.3676	0.3719		
Sulphurous acid			0.3845	0.3903		
Cyanogen			0.3829	0.3877		

GASES.

Centi-metres.	Millimetres.				
	0	1	2	3	4
0		7.11919	7.42022	7.59631	7.72125
1	8.11919	8.16058	8.19837	8.23313	8.26531
2	.42022	.44141	.46161	.48091	.49940
3	.59631	.61055	.62434	.63770	.65067
4	.72125	.73197	.74244	.75265	.76264
5	.81816	.82676	.83519	.84346	.85158
6	.89734	.90452	.91158	.91853	.92537
7	.96128	.97043	.97652	.98251	.98841
8	9.02228	9.02767	9.03300	9.03826	9.04346
9	.07343	.07823	.08297	.08769	.09231
10	.11919	.12351	.12779	.13202	.13622
11	.16058	.16451	.16840	.17226	.17609
12	.19837	.20197	.20555	.20909	.21261
13	.23313	.23646	.23976	.24304	.24629
14	.26531	.26841	.27147	.27452	·.27755
15	.29528	.29816	.30103	.30388	.30671
16	.32331	.32601	.32870	.33137	.33403
17	.34964	.35218	.35471	.35723	.35974
18	.37446	.37687	.37926	.38164	.38400
19	.39794	.40022	.40249	.40474	.40699
20	.42022	.42238	.42454	.42668	.42882
21	.44141	.44347	.44552	.44757	.44960
22	.46161	.46358	.46554	.46749	.46943
23	.48091	.48280	.48467	.48654	.48840
24	.49940	.50120	.50300	.50479	.50658
25	.51713	.51886	.52059	.52231	.52402
26	.53416	.53583	.53749	.53914	.54079
27	.55055	.55216	.55376	.55535	.55694
28	.56634	.56789	.56944	.57097	.57250
29	.58158	.58308	.58457	.58605	.58753
30	.59631	.59775	.59919	.60063	.60206
31	.61055	.61195	.61334	.61473	.61612
32	.62434	.62569	.62704	.62839	.62973
33	.63770	.63901	.64032	.64163	.64293
34	.65067	.65194	.65321	.65448	.65574

FROM $h = 1$ TO $h = 699$.

Centi-metres.	Millimetres.				
	5	6	7	8	9
0	7.81816	7.89734	7.96428	8.02228	8.07343
1	8.29538	8.32331	8.34964	.37446	.39794
2	.51713	.53416	.55055	.56634	.58158
3	.66325	.67549	.68739	.69897	.71025
4	.77240	.78194	.79128	.80043	.80938
5	.85956	.86737	.87506	.88261	.89004
6	.93210	.93873	.94526	.95169	.95804
7	.99425	9.00000	9.00568	9.01128	9.01681
8	9.04860	.05368	.05871	.06367	.06858
9	.09691	.10145	.10596	.11041	.11482
10	.14038	.14449	.14857	.15261	.15661
11	.17988	.18364	.18737	.19107 ·	.19473
12	.21610	.21956	.22299	.22640	.22978
13	.24952	.25273	.25591	.25907	.26220
14	.28055	.28354	.28650	.28945	.29237
15	.30952	.31231	.31509	.31784	.32058
16	.33667	.33929	.34190	.34450	.34707
17	.36222	.36470	.36716	.36961	.37204
18	.38636	.38870	.39103	.39334	.39565
19	.40922	.41144	.41365	.41585	.41804
20	.43094	.43305	.43516	.43725	.43933
21	.45162	.45364	.45565	.45764	.45963
22	.47137	.47329	.47521	.47712	.47902
23	.49025	.49210	.49393	.49576	.49758
24	.50835	.51012	.51188	.51364	.51539
25	.52573	.52743	.52912	.53081	.53249
26	.54243	.54407	.54570	.54732	.54894
27	.55852	.56010	.56167	.56323	.56479
28	.57403	.57555	.57707	.57858	.58008
29	.58901	.59048	.59194	.59340	.59486
30	.60349	.60491	.60632	.60774	.60914
31	.61750	.61887	.62025	.62161	.62298
32	.63107	.63240	.63373	.63506	.63638
33	.64423	.64553	.64682	.64810	.64939
34	.65701	.65826	.65952	.66077	.66201

GASES.

Centi-metres.	Millimetres.				
	0	1	2	3	4
35	9.66325	9.66449	9.66573	9.66696	9.66819
36	.67549	.67669	.67790	.67909	.68029
37	.68739	.68856	.68973	.69090	.69206
38	.69897	.70011	.70125	.70239	.70352
39	.71025	.71136	.71247	.71358	.71468
40	.72125	.72233	.72341	.72449	.72556
41	.73197	.73303	.73408	.73514	.73619
42	.74244	.74347	.74450	.74553	.74655
43	.75265	.75366	.75467	.75567	.75668
44	.76264	.76363	.76461	.76559	.76657
45	.77240	.77336	.77432	.77528	.77624
46	.78194	.78289	.78383	.78477	.78570
47	.79128	.79221	.79313	.79405	.79496
48	.80043	.80133	.80223	.80313	.80403
49	.80938	.81027	.81115	.81203	.81291
50	.81816	.81902	.81989	.82075	.82162
51	.82676	.82761	.82846	.82930	.83015
52	.83519	.83602	.83686	.83769	.83852
53	.84346	.84428	.84510	.84591	.84673
54	.85158	.85238	.85319	.85399	.85479
55	.85955	.86034	.86113	.86191	.86270
56	.86737	.86815	.86892	.86969	.87047
57	.87506	.87582	.87658	.87734	.87810
58	.88261	.88336	.88411	.88486	.88560
59	.89004	.89077	.89151	.89224	.89297
60	.89734	.89806	.89878	.89950	.90022
61	.90452	.90523	.90594	.90665	.90735
62	.91158	.91228	.91298	.91367	.91437
63	.91853	.91922	.91990	.92059	.92128
64	.92537	.92604	.92672	.92740	.92807
65	.93210	.93277	.93343	.93410	.93476
66	.93873	.93939	.94004	.94070	.94135
67	.94526	.94591	.94656	.94720	.94785
68	.95170	.95233	.95297	.95361	.35424
69	.95804	.95866	.95929	.95992	.96055

Centi-metres.	Millimetres.				
	5	6	7	8	9
35	9.66941	9.67064	9.67185	9.67307	9.67428
36	.68148	.68267	.68385	.68503	.68621
37	.69322	.69437	.69553	.69668	.69783
38	.70465	.70577	.70690	.70802	.70914
39	.71578	.71688	.71798	.71907	.72016
40	.72664	.72771	.72878	.72985	.73091
41	.73723	.73828	.73932	.74036	.74140
42	.74758	.74860	.74961	.75063	.75164
43	.75768	.75867	.75967	.76066	.76165
44	.76755	.76852	.76949	.77046	.77143
45	.77720	.77815	.77910	.78005	.78100
46	.78664	.78757	.78850	.78943	.79036
47	.79588	.79679	.79770	.79861	.79952
48	.80493	.80582	.80672	.80760	.80850
49	.81379	.81467	.81554	.81642	.81729
50	.82248	.82334	.82419	.82505	.82590
51	.83099	.83184	.83268	.83352	.83435
52	.83935	.84017	.84100	.84182	.84264
53	.84754	.84835	.84916	.84997	.85078
54	.85558	.85638	.85717	.85797	.85876
55	.86348	.86426	.86504	.86582	.86660
56	.87123	.87200	.87277	.87353	.87430
57	.87885	.87961	.88036	.88111	.81870
58	.88634	.88708	.88782	.88856	.88930
59	.89370	.89443	.89516	.89589	.89661
60	.90094	.90166	.90238	.90309	.90380
61	.90806	.90877	.90947	.91017	.91088
62	.91507	.91576	.91645	.91715	.91784
63	.92196	.92264	.92333	.92401	.92469
64	.92875	.92942	.93009	.93076	.93143
65	.93543	.93609	.93675	.93741	.93807
66	.94201	.94266	.94331	.94396	.94461
67	.94849	.94913	.94978	.95042	.95106
68	.95488	.95551	.95614	.95677	.95741
69	.96117	.96180	.96242	.96304	.96366

BAROMETRICAL CORRECTION. LOGARITHMS OF $\frac{h}{760}$.

Milli-metres.	Tenths of Millimetres.									
	0.0	0.1	0.2	0.3	0.4	0.5	0.6	0.7	0.8	0.9
700	9.96428	434	441	447	453	459	465	472	478	484
701	490	496	503	509	515	521	527	534	540	546
702	552	558	565	571	577	583	589	596	602	608
703	614	620	626	632	639	645	651	657	663	670
704	676	682	688	694	701	707	713	719	725	731
705	737	743	749	755	762	768	774	780	786	793
706	799	805	811	817	823	829	836	842	848	854
707	860	866	872	878	884	890	896	903	909	915
708	922	928	934	940	946	952	958	964	970	976
709	982	988	994	*001	*007	*013	*019	*025	*021	*038
710	9.97044	050	056	062	068	074	081	087	093	099
711	105	111	117	124	130	136	142	148	154	160
712	166	172	178	185	191	197	203	209	215	221
713	227	233	239	246	252	258	264	270	276	282
714	288	294	300	307	313	319	325	331	337	343
715	349	355	361	367	373	379	385	391	397	404
716	410	416	422	428	434	440	446	452	458	464
717	470	476	482	488	494	500	506	513	519	525
718	531	537	543	549	555	561	567	573	579	585
719	591	597	603	609	615	621	627	633	639	645
720	651	657	663	669	675	681	687	693	699	705
721	711	717	723	729	735	742	748	754	760	766
722	772	778	784	790	796	802	808	814	820	826
723	832	838	844	850	856	862	868	874	880	886
724	892	898	904	910	916	922	928	934	940	946
725	952	958	964	970	976	982	988	994	*000	*006
726	9.98012	018	024	030	036	042	048	054	060	066
727	072	078	084	090	096	102	108	114	120	126
728	132	138	144	150	156	162	168	173	179	185
729	191	197	203	209	215	221	227	233	239	245
730	251	257	263	269	275	280	286	292	298	304
731	310	316	322	328	334	340	346	352	358	364
732	370	376	381	387	393	399	405	411	417	423
733	429	435	441	447	453	458	464	470	476	482
734	488	494	500	506	512	518	524	530	535	541
735	547	553	559	565	571	577	583	589	595	601
736	607	612	618	624	630	636	642	648	653	659
737	665	671	677	683	689	695	701	706	712	718
738	724	730	736	742	748	754	760	765	771	777
739	783	789	795	801	806	812	818	824	830	836

Milli-metres.	Tenths of Millimetres.									
	0.0	0.1	0.2	0.3	0.4	0.5	0.6	0.7	0.8	0.9
740	9.98842	848	854	860	866	872	877	883	889	895
741	901	907	913	918	924	930	936	942	948	954
742	959	965	971	977	983	989	994	*000	*006	*012
743	9.99018	024	030	035	041	047	053	059	065	070
744	076	082	088	094	100	105	111	117	123	129
745	135	140	146	152	158	164	170	175	181	187
746	193	199	205	210	216	222	228	234	239	245
747	251	257	263	268	274	280	286	292	298	303
748	309	315	321	327	332	338	344	350	356	361
749	367	373	379	385	390	396	402	408	414	419
750	425	431	437	442	448	454	460	466	471	477
751	483	489	495	500	506	512	518	523	529	535
752	541	547	552	558	564	570	575	581	587	593
753	598	604	610	616	622	627	633	639	645	650
754	656	662	668	673	679	685	691	696	702	708
755	714	719	725	731	737	742	748	754	760	765
756	771	777	783	788	794	800	806	811	817	823
757	829	834	840	846	852	857	863	869	874	880
758	886	892	897	903	909	915	920	926	932	937
759	943	949	955	960	966	972	977	983	989	995
760	0.00000	006	012	017	023	029	035	040	046	052
761	057	063	069	075	080	086	092	097	103	109
762	114	120	126	132	137	143	149	154	160	166
763	171	177	183	189	194	200	206	211	217	223
764	228	234	240	245	251	257	262	268	274	279
765	285	291	296	302	308	314	319	325	331	336
766	342	348	353	359	365	370	376	382	387	393
767	399	404	410	416	421	427	432	438	444	449
768	455	461	466	472	478	483	489	495	500	506
769	512	517	523	529	534	540	546	551	557	562
770	568	573	579	584	590	596	601	607	612	618
771	624	629	635	641	646	652	657	663	669	674
772	680	686	691	697	703	708	714	719	725	731
773	736	742	747	753	758	764	770	776	781	787
774	793	798	804	809	815	821	826	832	837	843
775	849	854	860	865	871	877	882	888	894	899
776	905	910	916	921	927	934	938	944	949	955
777	961	966	972	978	983	989	994	*000	*006	*011
778	0.01017	022	028	033	039	044	050	056	061	067
779	072	078	083	089	095	100	106	111	117	123
780	128	134	139	145	150	156	162	167	173	178

TABLE OF LOG $\dfrac{1}{1+0.00367t}$.

t		Tenths of Degrees.								
	0.0	0.1	0.2	0.3	0.4	0.5	0.6	0.7	0.8	0.9
0		984	968	952	936	920	904	888	872	857
1	9.99841	825	809	793	777	762	746	730	714	698
2	683	667	651	635	619	604	588	572	556	541
3	525	509	493	478	462	446	430	415	399	383
4	368	352	336	321	305	289	274	258	242	227
5	211	196	180	164	149	133	117	102	086	071
6	055	040	024	008	*993	*977	*962	*946	*931	*915
7	9.98900	884	869	853	838	822	807	791	776	760
8	745	729	714	698	683	668	652	637	621	606
9	590	575	560	544	529	514	498	483	467	452
10	437	421	406	391	375	360	345	329	314	299
11	283	268	253	238	222	207	192	177	161	146
12	131	116	100	085	070	055	039	024	009	*994
13	9.97979	963	948	933	918	903	888	872	857	842
14	827	812	797	782	766	751	736	721	706	691
15	676	661	646	631	616	600	585	570	555	540
16	525	510	495	480	465	450	435	420	405	390
17	375	360	345	330	315	300	285	270	255	241
18	226	211	196	181	166	151	136	121	106	091
19	076	062	047	032	017	002	*987	*972	*958	*943
20	9.96928	913	898	883	869	854	839	824	809	795
21	780	765	750	735	721	706	691	676	662	647
22	632	618	603	588	573	559	544	529	515	500
23	485	471	456	441	426	412	397	383	368	353
24	339	324	309	295	280	266	251	237	222	207
25	188	174	159	145	130	116	101	087	072	057
26	042	028	013	*999	*984	*970	*955	*941	*926	*912
27	9.95897	883	868	854	839	825	810	796	781	767
28	752	738	723	709	694	680	665	651	636	622
29	607	593	578	564	549	535	520	506	491	477
30	463	449	434	420	406	391	377	363	348	334
31	320	306	292	277	263	249	234	220	206	191
32	177	163	149	134	120	106	092	078	063	049
33	035	021	007	*993	*978	*964	*950	*935	*921	*907
34	9.94893	879	865	851	837	822	808	794	780	766

t	0.0	0.1	0.2	0.3	0.4	0.5	0.6	0.7	0.8	0.9
					Tenths of Degrees.					
35°	9.94752	738	724	710	695	681	667	653	639	625
36	611	597	583	569	554	540	526	512	498	484
37	470	456	442	428	414	400	386	372	358	344
38	330	316	302	288	274	260	246	232	218	204
39	190	176	162	148	134	121	107	093	079	065
40	051	037	023	009	*996	*982	*968	*954	*940	*926
41	9.93912	898	884	871	857	843	829	816	802	788
42	774	760	746	733	719	705	692	678	665	651
43	637	623	609	595	582	568	554	540	527	513
44	499	485	471	458	444	430	417	403	390	376
45	362	348	334	321	307	293	279	266	253	239
46	225	211	198	184	171	158	144	131	117	103
47	089	075	062	048	035	021	007	*994	*980	*967
48	9.92953	940	926	913	899	885	871	858	844	831
49	818	804	791	777	764	750	737	723	710	696
50	683	670	656	643	629	616	603	589	576	562
51	549	536	522	509	496	482	469	455	442	428
52	415	402	388	375	361	348	335	321	308	294
53	281	268	254	240	227	214	200	187	173	160
54	147	134	120	107	094	080	067	054	040	027
55	014	001	*988	*974	*961	*948	*935	*921	*908	*895
56	9.91882	869	856	842	829	816	803	790	776	763
57	750	737	724	711	698	684	671	658	645	632
58	619	606	593	580	567	553	540	527	514	501
59	488	475	462	449	436	422	409	396	383	370
60	357	344	331	318	305	292	279	266	253	240
61	227	214	201	188	175	162	149	136	123	110
62	097	084	071	058	045	032	019	006	*993	*980
63	9.90967	954	941	928	915	902	889	876	863	851
64	838	825	812	799	786	773	760	747	735	722
65	709	696	683	670	657	644	632	619	606	593
66	580	567	554	542	529	516	503	490	478	465
67	452	439	426	414	401	388	375	363	350	337
68	324	311	298	286	273	260	248	235	223	210
69	197	184	172	159	146	134	121	108	096	083
70	070	057	044	032	019	006	*993	*980	*968	*956
71	9.89943	930	918	905	893	880	867	855	842	830
72	817	805	792	780	767	755	742	730	717	704
73	691	679	666	654	641	628	616	603	591	578
74	565	553	540	528	515	503	490	478	465	453

t	Tenths of Degrees.									
	0.0	**0.1**	**0.2**	**0.3**	**0.4**	**0.5**	**0.6**	**0.7**	**0.8**	**0.9**
75	9.89440	428	415	403	390	378	366	353	341	328
76	316	304	291	279	266	254	241	229	216	204
77	91	179	166	154	141	129	117	104	092	079
78	067	054	042	029	017	005	*992	*980	*967	*955
79	9.88943	931	919	906	894	882	869	857	845	833
80	820	808	796	783	771	759	747	734	722	710
81	697	685	673	661	648	636	624	612	599	587
82	574	562	550	533	525	513	501	489	477	464
83	452	440	428	416	403	391	379	367	354	342
84	330	318	306	294	282	269	257	245	233	221
85	209	197	185	173	160	148	136	124	111	099
86	087	075	063	051	039	026	014	002	*990	*978
87	9.87966	954	942	930	918	905	893	881	869	857
88	845	833	821	809	797	785	773	761	749	737
89	725	713	701	689	677	665	653	641	629	617
90	605	593	581	569	557	545	533	521	509	497
91	485	473	461	449	437	426	414	402	390	378
92	366	354	342	330	318	306	295	283	271	259
93	247	235	223	212	200	188	176	165	153	141
94	129	117	105	093	081	070	058	046	034	022
95	010	*998	*986	*975	*963	*951	*939	*928	*916	*904
96	9.86892	880	868	857	845	833	821	809	798	786
97	774	762	750	739	727	715	704	692	680	669
98	657	645	633	621	610	598	586	574	563	551
99	539	527	516	504	493	481	469	458	446	435
100	423	412	400	389	378	366	355	343	331	319
101	307	296	284	273	261	249	238	226	214	203
102	191	179	168	156	145	133	121	110	098	087
103	075	063	052	040	029	017	005	*994	*982	*971
104	9.85959	948	936	925	913	902	890	879	867	856
105	844	833	821	810	798	787	775	764	752	741
106	729	718	706	695	683	672	661	649	638	626
107	615	603	592	580	569	557	546	534	523	511
108	500	488	477	465	454	443	431	420	408	397
109	386	374	363	351	340	329	317	306	295	283
110	272	261	249	238	227	215	204	193	181	170
111	159	148	136	125	114	102	091	080	069	057
112	046	035	023	012	001	*989	*978	*967	*955	*944
113	9.84933	922	911	899	888	877	866	854	843	832
114	821	810	799	787	776	765	754	742	731	720

t	Tenths of Degrees.									
	0.0	0.1	0.2	0.3	0.4	0.5	0.6	0.7	0.8	0.9
115	9.84709	698	687	675	664	653	642	630	619	608
116	597	586	575	563	552	541	530	518	507	496
117	485	474	463	451	440	429	418	407	395	384
118	373	362	351	340	329	317	306	295	284	273
119	262	251	240	229	218	207	196	185	174	163
120	152	141	130	119	107	096	085	074	063	052
121	041	030	019	008	*997	*985	*974	*963	*952	*941
122	9.83931	920	909	898	887	876	865	854	843	832
123	821	810	799	788	777	766	756	745	734	723
124	712	701	690	679	668	657	646	635	624	613
125	602	591	580	569	558	547	536	525	514	503
126	493	482	471	460	449	438	428	417	406	395
127	384	373	362	351	340	329	319	308	297	286
128	275	264	253	243	232	221	210	200	189	178
129	167	156	145	135	124	113	102	091	081	070
130	059	048	037	027	016	005	*994	*984	*973	*962
131	9.82951	940	929	919	908	897	886	876	865	854
132	844	833	822	812	801	790	779	769	758	747
133	737	726	715	705	694	683	672	662	651	640
134	630	619	608	598	587	576	565	554	544	533
135	523	512	501	491	480	469	458	448	437	426
136	416	405	395	384	373	363	352	342	331	320
137	310	299	289	278	267	257	246	235	225	214
138	204	193	183	172	162	151	141	130	120	109
139	099	088	078	067	057	046	035	025	014	004
140	9.81993	983	972	961	951	940	929	919	908	898
141	888	877	867	856	846	835	825	814	804	793
142	783	772	762	751	741	730	720	709	699	688
143	678	667	657	646	636	626	615	605	594	584
144	574	564	553	543	533	522	512	501	491	480
145	470	460	449	439	428	418	408	397	387	376
146	366	356	345	335	324	314	303	293	283	272
147	262	252	241	231	220	210	200	189	179	168
148	158	148	138	127	117	107	096	086	075	065
149	055	045	035	025	014	004	*994	*984	*973	*963
150	9.80953	943	932	922	911	901	891	881	871	860
151	850	840	830	819	809	799	789	779	768	758
152	748	738	728	717	707	697	687	677	667	656
153	646	636	626	615	605	595	585	575	565	554
154	544	534	524	514	503	493	483	473	463	452

t	0.0	0.1	0.2	0.3	0.4	0.5	0.6	0.7	0.8	0.9
					Tenths of Degrees.					
155	9.80442	432	422	412	402	391	381·	471	361	351
156	340	330	320	310	300	289	279	269	259	249
157	239	229	219	209	199	189	178	168	158	148
158	138	128	118	108	098	088	078	067	057	047
159	037	027	017	007	*997	*987	*977	*967	*957	*947
160	9.79937	927	917	907	897	887	877	867	857	847
161	837	827	817	807	797	787	777	767	757	747
162	737	727	717	707	697	687	677·	667	657	647
163	637	627	617	607	597	587	577	567	557	547
164	537	527	517	507	497	487	477	467	457	447
165	437	427	417	407	397	388	378	368	358	348
166	338	328	·318	309	299	289	279	270	260	250
167	240	230	220	210	200	191	181	171	161	151
168	141	131	121	111	102	092	082	072	062	052
169	042	032	022	013	003	*993	*983	*973	*964	*954
170	9.78944	934	925	915	905	895	885	876	866	856
171	846	836	827	817	807	797	787	777	768	758
172	748	738	729	719	709	699	689	680	670	660
173	650	640	731	621	611	602	592	582	572	563
174	553	543	533	524	514	504	494	485	475	466
175	456	446	437	427	417	408	398	389	379	370
176	360	350	341	331	321	312	302	292	283	273
177	263	253	244	234	224	215	205	195	185	176
178	166	156	147	137	128	118	108	099	089	080
179	070	060	051	041	031	022	012	002	*993	*983
180	9.77973	964	954	945	935	926	916	906	897	887
181	878	868	859	849	840	830	821	811	801	792
182	782	772	763	753	743	734	724	715	705	696
183	686	677	667	658	648	639	629	620	610	601
184	591	582	572	563	553	544	534	525	515	506
185	496	487	477	468	458	449	439	430	420	411
186	402	392	383	373	364	354	345	335	326	316
187	307	297	288	278·	269	259	250	240	231	221
188	212	202	193	183	174	164	155	146	136	127
189	118	108	099	089	080	071	061	052	043	033
190	024	014	005	*996	*986	*977	*968	*958	*949	*940
191	9.76931	921	912	903	893	884	875	865	856	847
192	837	828	818	809	799	790	781	771	762	753
193	743	734	725	715	706	697	687	678	669	659
194	650	640	631	622	612	603	594	585	575	566

t	0.0	0.1	0.2	0.3	0.4	0.5	0.6	0.7	0.8	0.9
					Tenths of Degrees.					
195	9.76557	547	538	529	519	510	501	492	482	473
196	464	455	445	436	427	417	408	399	390	380
197	371	362	352	343	334	325	316	306	297	288
198	279	270	260	251	242	233	224	215	205	196
199	187	178	169	159	150	141	132	123	113	104
200	095	086	077	067	058	049	040	031	021	012
201	003	*994	*985	*975	*966	*957	*948	*939	*930	*920
202	9.75911	902	893	884	875	865	856	847	838	829
203	820	811	802	793	784	774	765	756	747	738
204	729	720	711	702	693	683	674	665	656	647
205	638	629	620	611	602	592	583	573	565	556
206	547	538	529	520	511	501	492	483	474	465
207	456	447	438	429	420	411	402	393	384	375
208	366	357	348	339	330	320	311	302	293	284
209	275	266	257	248	239	230	221	212	203	194
210	185	176	167	158	149	140	131	122	113	104
211	095	086	077	068	059	050	042	033	024	015
212	006	*997	*988	*979	*970	*961	*953	*944	*935	*926
213	9.74917	908	899	890	881	872	863	854	845	836
214	827	818	809	800	791	782	774	765	756	747
215	738	729	720	711	702	693	685	676	667	658
216	649	640	631	622	613	604	596	587	578	569
217	560	551	542	533	524	515	507	498	489	480
218	471	462	453	444	435	426	418	409	400	391
219	382	373	364	356	347	338	329	320	312	303
220	294	285	276	268	259	250	241	232	224	215
221	206	197	188	180	171	162	153	145	136	127
222	118	109	100	092	083	074	066	057	048	040
223	031	022	013	005	*996	*987	*978	*969	*961	*952
224	9.73943	934	926	917	908	900	891	882	874	865
225	856	847	839	830	821	813	804	795	787	778
226	769	760	752	743	734	726	717	708	700	691
227	682	673	665	656	647	639	630	621	613	604
228	595	586	578	569	560	552	543	534	526	517
229	508	499	491	482	474	465	456	447	439	430
230	422	413	405	396	387	379	370	361	353	344
231	335	326	318	309	301	292	283	275	266	258
232	249	240	232	223	215	206	197	189	180	172
233	163	154	146	137	129	120	111	103	094	086
234	077	068	060	051	043	034	026	017	009	000

t	0.0	0.1	0.2	0.3	0.4	0.5	0.6	0.7	0.8	0.9
					Tenths of Degrees.					
235	9.72992	983	975	966	958	949	940	932	923	915
236	906	898	889	881	873	864	856	847	839	830
237	821	813	804	796	787	779	770	762	753	745
238	736	728	719	711	702	694	685	677	668	660
239	651	643	634	626	617	609	600	592	583	575
240	566	558	549	541	532	524	515	507	498	490
241	481	473	464	456	447	439	431	422	414	405
242	397	388	380	371	363	354	346	338	329	321
243	313	304	296	287	279	271	262	254	245	237
244	229	221	212	204	195	187	179	170	162	153
245	145	137	128	120	111	103	095	086	078	069
246	061	053	044	036	027	019	011	002	*994	*985
247	9.71977	969	960	952	943	935	927	918	910	901
248	893	885	876	868	858	851	843	834	826	818
249	810	802	793	785	777	768	760	752	743	735
250	726	718	709	701	693	684	676	668	659	651
251	643	635	626	618	610	601	593	585	576	568
252	560	552	544	535	527	519	511	503	494	486
253	478	470	461	453	445	437	428	420	412	403
254	395	387	379	370	362	354	346	338	329	321
255	313	305	297	288	280	272	264	255	247	239
256	231	223	215	206	198	190	182	174	165	157
257	149	141	133	124	116	108	100	092	083	075
258	067	059	051	042	034	026	018	010	001	*993
259	9.70985	977	969	960	952	944	936	928	919	911
260	903	895	887	878	870	862	854	846	838	830
261	822	814	806	797	789	781	773	765	756	748
262	740	732	724	716	708	699	691	683	675	667
263	659	651	643	635	627	618	610	602	594	586
264	578	570	562	554	546	538	530	521	513	505
265	497	489	481	473	465	457	448	440	432	424
266	416	408	400	392	384	375	367	359	351	343
267	335	327	319	311	303	295	287	279	271	263
268	255	247	239	231	223	215	207	199	191	183
269	175	167	159	151	143	135	127	119	111	103
270	095	087	079	071	063	055	047	039	031	023
271	015	007	*999	*991	*983	*975	*967	*959	*951	*943
272	9.69935	927	919	911	903	895	888	880	872	864
273	856	848	840	832	824	816	808	800	792	784
274	776	768	760	752	744	736	728	720	712	704

t	Tenths of Degrees.									
	0.0	0.1	0.2	0.3	0.4	0.5	0.6	0.7	0.8	0.9
275	9.69696	688	680	672	664	656	648	640	632	625
276	617	609	601	593	585	577	570	562	554	546
277	538	530	522	514	506	499	491	483	475	467
278	459	451	443	436	428	420	412	404	397	389
279	381	373	365	357	349	341	334	326	318	310
280	302	294	286	279	271	263	255	248	240	232
281	224	216	208	200	192	185	177	169	161	153
282	145	137	129	122	114	106	098	090	083	075
283	067	059	051	043	035	027	020	012	004	*996
284	3.68988	980	972	965	957	949	941	933	926	918
285	910	902	895	887	879	871	864	856	848	841
286	833	825	817	810	802	794	786	778	771	763
287	755	747	740	732	724	716	709	701	693	686
288	678	670	662	655	647	639	631	623	616	608
289	600	592	585	577	569	561	554	546	538	531
290	523	515	508	500	492	484	477	469	461	454
291	446	438	431	423	415	407	399	392	384	377
292	369	361	354	346	338	330	323	315	307	300
293	292	284	277	269	261	253	246	238	230	223
294	215	207	200	192	184	176	169	161	153	146
295	138	130	123	115	107	100	092	085	077	070
296	062	054	047	039	031	023	016	008	000	*993
297	9.67985	977	970	962	954	947	939	932	924	917
298	909	901	894	886	878	871	863	856	848	840
299	833	825	818	810	803	795	787	780	773	765
300	757	749	742	734	627	719	711	704	696	689
301	681	673	666	658	651	643	636	628	621	613
302	606	598	591	583	576	568	561	553	546	538
303	531	523	516	508	501	493	485	478	470	463
304	455	448	440	433	425	418	410	403	395	388
305	380	373	365	358	350	343	335	328	320	313
306	305	298	290	283	275	268	260	253	245	238
307	230	223	215	208	200	193	185	178	170	163
308	155	148	140	133	125	118	110	103	095	088
309	030	073	065	058	050	043	035	028	020	013
310	005	*998	*990	*983	*975	*968	*961	*953	*946	*938
311	9.66931	923	916	908	901	893	886	878	871	863
312	856	848	841	834	826	819	812	804	797	789
313	782	774	767	759	752	744	737	730	722	715
314	708	700	693	685	678	671	663	656	648	641

t	Tenths of Degrees.									
	0.0	0.1	0.2	0.3	0.4	0.5	0.6	0.7	0.8	0.9
315	9.66634	626	619	612	604	597	589	582	574	567
316	560	552	545	537	530	523	515	508	500	493
317	486	478	471	464	456	449	441	434	427	420
318	413	405	398	391	383	376	369	361	354	347
319	339	332	324	317	310	302	295	288	280	273
320	266	259	251	244	237	229	222	215	207	200
321	193	186	178	171	164	156	149	142	134	127
322	120	113	105	098	091	083	076	069	061	054
323	047	040	032	025	018	010	003	*996	*988	*981
324	9.65974	967	959	952	945	937	930	923	915	908
325	901	894	886	879	872	864	857	850	842	835
326	828	821	814	806	799	792	785	777	770	763
327	756	749	741	734	727	720	713	705	698	691
328	684	677	669	662	655	647	640	633	625	618
329	611	604	596	589	582	575	568	560	553	546
330	539	532	524	517	510	503	496	488	481	474
331	467	460	452	445	438	431	424	417	409	402
332	395	388	381	374	367	359	352	345	338	331
333	324	317	309	302	295	288	281	273	266	259
334	252	245	237	230	223	216	209	201	194	187
335	180	173	166	159	152	144	137	130	123	116
336	109	102	094	087	080	073	066	058	051	044
337	037	030	023	016	009	001	*994	*987	*980	*973
338	9.64966	959	952	945	938	930	923	916	909	902
339	895	888	881	874	867	859	852	845	838	831
340	824	817	810	803	796	788	781	774	767	760
341	753	746	739	732	725	718	710	703	696	689
342	682	675	668	661	654	647	640	633	626	619
343	612	605	598	591	584	577	570	563	556	549
344	542	535	528	521	514	507	500	493	486	479
345	471	464	457	450	443	436	429	422	415	408
346	401	394	387	380	373	366	359	352	345	338
347	331	324	317	310	303	296	289	282	275	268
348	261	254	247	240	233	225	218	211	204	197
349	190	183	176	169	162	155	148	141	134	127
350	120	113	106	099	092	085	078	071	064	057

TABLE

OF THE ELASTIC FORCE OF THE VAPOR OF WATER FROM −32° TO +230°.

REGNAULT. — Mémoires de l'Académie de France, Tome XXI. p. 624.

Temperature.	Elastic Force.	Difference for 1°.	Temperature.	Elastic Force.	Difference for 1°.	Temperature.	Elastic Force.	Difference for 1°.
°	mm.	mm.	°	mm.	mm.	°	mm.	mm.
−32	0.320	0.032	9	8.574	0.591	50	91.982	4.679
31	0.352	0.034	10	9.165	0.627	51	96.661	4.882
30	0.386	0.038	11	9.792	0.665	52	101.543	5.093
29	0.424	0.040	12	10.457	0.705	53	106.636	5.309
28	0.464	0.044	13	11.162	0.746	54	111.945	5.533
27	0.508	0.047	14	11.908	0.791	55	117.478	5.766
26	0.555	0.050	15	12.699	0.837	56	123.244	6.007
25	0.605	0.055	16	13.536	0.885	57	129.251	6.254
24	0.660	0.059	17	14.421	0.936	58	135.505	6.510
23	0.719	0.064	18	15.357	0.989	59	142.015	6.776
22	0.783	0.070	19	16.346	1.045	60	148.791	7.048
21	0.853	0.074	20	17.391	1.104	61	155.839	7.331
20	0.927	0.081	21	18.495	1.164	62	163.170	7.621
19	1.008	0.087	22	19.659	1.229	63	170.791	7.923
18	1.095	0.094	23	20.888	1.296	64	178.714	8.231
17	1.189	0.101	24	22.184	1.366	65	186.945	8.551
16	1.290	0.110	25	23.550	1.438	66	195.496	8.880
15	1.400	0.118	26	24.988	1.517	67	204.376	9.220
14	1.518	0.128	27	26.505	1.596	68	213.596	9.569
13	1.646	0.137	28	28.101	1.681	69	223.165	9.928
12	1.783	0.150	29	29.782	1.766	70	233.093	10.300
11	1.933	0.160	30	31.548	1.858	71	243.393	10.680
10	2.093	0.174	31	33.406	1.953	72	254.073	11.074
9	2.267	0.188	32	35.359	2.052	73	265.147	11.477
8	2.455	0.203	33	37.411	2.154	74	276.624	11.893
7	2.658	0.218	34	39.565	2.262	75	288.517	12.321
6	2.876	0.237	35	41.827	2.374	76	300.838	12.762
5	3.113	0.245	36	44.201	2.490	77	313.600	13.211
4	3.368	0.276	37	46.691	2.611	78	326.811	13.677
3	3.644	0.297	38	49.302	2.737	79	340.488	14.155
2	3.941	0.322	39	52.039	2.867	80	354.643	14.644
− 1	4.263	0.337	40	54.906	3.004	81	369.287	15.148
0	4.600	0.340	41	57.910	3.145	82	384.435	15.666
+ 1	4.940	0.362	42	61.055	3.291	83	400.101	16.197
2	5.302	0.385	43	64.346	3.444	84	416.298	16.743
3	5.687	0.410	44	67.790	3.601	85	433.041	17.303
4	6.097	0.437	45	71.391	3.767	86	450.344	17.877
5	6.534	0.464	46	75.158	3.935	87	468.221	18.466
6	6.998	0.494	47	79.093	4.111	88	486.687	19.072
7	7.492	0.525	48	83.204	4.295	89	505.759	19.691
8	8.017	0.557	49	87.499	4.483	90	525.450	20.328

Temperature.	Elastic Force.	Difference for 1°.	Temperature.	Elastic Force.	Difference for 1°.	Temperature.	Elastic Force.	Difference for 1°.
°	mm.	mm.	°	mm.	mm.	°	mm.	mm.
91	545.778	20.979	138	2567.000	77.440	185	8453.23	191.12
92	566.757	21.649	139	2641.440	76.190	186	8644.35	194.47
93	588.406	22.334	140	2717.630	77.940	187	8838.82	197.86
94	610.740	23.038	141	2795.570	79.730	188	9036.68	201.27
95	633.778	23.757	142	2875.300	81.560	189	9237.95	204.75
96	657.535	24.494	143	2956.860	83.400	190	9442.70	208.23
97	682.029	25.251	144	3040.260	85.290	191	9650.93	211.78
98	707.280	26.025	145	3125.550	87.190	192	9862.71	215.33
99	733.305	26.695	146	3212.740	89.130	193	10078.04	218.97
100	760.000	27.590	147	3301.870	91.110	194	10297.01	222.62
101	787.590	28.420	148	3392.980	93.110	195	10519.63	226.32
102	816.010	29.270	149	3486.090	95.140	196	10745.95	230.05
103	845.280	30.130	150	3581.230	97.200	197	10975.00	233.82
104	875.410	31.000	151	3678.430	99.310	198	11209.82	237.64
105	906.410	31.900	152	3777.740	101.440	199	11447.46	241.50
106	938.310	32.830	153	3879.180	103.590	200	11688.96	245.41
107	971.140	33.770	154	3982.770	105.790	201	11934.37	249.32
108	1004.910	34.740	155	4088.560	108.030	202	12183.69	253.31
109	1039.650	35.720	156	4196.590	110.290	203	12437.00	257.30
110	1075.370	36.720	157	4306.880	112.570	204	12694.30	261.36
111	1112.090	37.740	158	4419.450	114.910	205	12955.66	265.46
112	1149.830	38.780	159	4534.860	117.260	206	13221.12	269.63
113	1188.610	39.860	160	4651.620	119.660	207	13490.75	273.78
114	1228.470	40.940	161	4771.280	122.080	208	13764.53	277.99
115	1269.410	42.060	162	4893.360	124.550	209	14042.52	282.28
116	1311.470	43.190	163	5017.910	127.060	210	14324.80	286.52
117	1354.660	44.340	164	5144.970	129.570	211	14611.32	290.90
118	1399.020	45.530	165	5274.540	132.150	212	14902.22	295.26
119	1444.550	46.730	166	5406.690	134.740	213	15197.48	299.69
120	1491.280	47.970	167	5541.430	137.390	214	15497.17	304.16
121	1539.250	49.220	168	5678.820	140.080	215	15801.33	308.61
122	1588.470	50.490	169	5818.900	142.760	216	16109.94	313.21
123	1638.960	51.800	170	5961.660	145.530	217	16423.15	317.75
124	1690.760	53.120	171	6107.190	148.290	218	16740.90	322.39
125	1743.880	54.470	172	6255.480	151.120	219	17063.29	327.07
126	1798.350	55.850	173	6406.600	153.950	220	17390.36	331.77
127	1854.200	57.270	174	6560.550	156.850	221	17722.13	336.55
128	1911.460	58.680	175	6717.430	159.790	222	18058.64	341.30
129	1970.150	60.130	176	6877.220	162.750	223	18399.94	346.13
130	2030.280	61.660	177	7039.970	165.750	224	18746.07	350.97
131	2091.940	63.090	178	7205.720	168.800	225	19097.04	355.88
132	2155.030	64.660	179	7374.520	171.870	226	19452.92	360.84
133	2219.690	66.230	180	7546.390	174.980	227	19813.76	365.85
134	2285.920	67.810	181	7721.370	178.150	228	20179.61	370.87
135	2353.730	69.430	182	7899.520	181.320	229	20550.48	375.92
136	2423.160	71.070	183	8080.840	184.560	230	20926.40	
137	2494.230	72.770	184	8265.400	187.830			

TABLE

OF THE TENSION OF THE VAPOR OF WATER IN MILLIMETRES FROM 85 TO 101 DEGREES.

REGNAULT. — Mémoires de l'Académie de France, Tome XXI. p. 632.

Degrees.	0.0	0.1	0.2	0.3	0.4	0.5	0.6	0.7	0.8	0.9
	mm.	mm.	mm.	mm.	mm.	mm.	mm.	mm.	mm.	mm.
85	433.04	434.75	435.46	438.17	439.89	441.62	443.35	445.09	446.84	448.59
86	450.34	452.10	453.87	455.64	457.42	459.21	461.00	462.80	464.60	466.41
87	468.22	470.04	471.87	473.70	475.54	477.38	479.23	481.08	482.94	484.81
88	486.69	488.57	490.45	492.34	494.24	496.15	498.06	499.98	501.90	503.82
89	505.76	507.70	509.65	511.60	513.56	515.53	517.50	519.48	521.46	523.45
90	525.45	527.45	529.46	531.48	533.50	535.53	537.57	539.61	541.66	543.72
91	545.78	547.85	549.92	552.00	554.09	556.19	558.29	560.39	562.51	564.63
92	566.76	568.89	571.03	573.18	575.34	577.50	579.67	581.84	584.02	586.20
93	588.41	590.61	592.82	595.04	597.26	599.49	601.72	603.97	606.22	608.48
94	610.74	613.01	615.29	617.58	619.85	622.17	624.48	626.79	629.11	631.44
95	633.78	636.12	638.47	640.83	643.19	645.57	647.95	650.34	652.73	655.13
96	657.54	659.95	662.37	664.80	667.24	669.69	672.14	674.60	677.07	679.55
97	682.03	684.52	687.02	689.53	692.04	694.56	697.08	699.61	702.15	704.70
98	707.26	709.82	712.39	714.97	717.56	720.15	722.75	725.35	727.96	730.58
99	733.21	735.85	738.50	741.16	743.83	746.50	749.18	751.87	754.57	757.28
100	760.00	762.73	765.46	768.20	771.95	773.71	776.48	779.26	782.04	784.83
101	787.63									

Tenths of Degrees.

TABLE

OF THE TENSION OF THE VAPOR OF WATER IN MILLIMETRES FOR EACH TENTH OF A DEGREE

FROM −10° TO +35°.

REGNAULT.—Mémoires de l'Académie de France, Tome XXI. p. 627.

Tenths of Degrees.

Degrees.	0.0	0.1	0.2	0.3	0.4	0.5	0.6	0.7	0.8	0.9
	mm.	mm.	mm.	mm.	mm.	mm.	mm.	mm.	mm.	mm.
−10	2.078									
9	2.261	2.242	2.223	2.204	2.186	2.168	2.150	2.132	2.114	2.096
8	2.456	2.436	2.416	2.396	2.376	2.356	2.337	2.318	2.299	2.280
7	2.666	2.645	2.624	2.603	2.582	2.561	2.540	2.519	2.498	2.477
6	2.890	2.867	2.844	2.821	2.798	2.776	2.754	2.732	2.710	2.688
5	3.131	3.106	3.082	3.058	3.034	3.010	2.986	2.962	2.938	2.914
4	3.387	3.361	3.335	3.309	3.283	3.257	3.231	3.206	3.181	3.156
3	3.662	3.634	3.606	3.578	3.550	3.522	3.495	3.468	3.441	3.414
2	3.955	3.925	3.895	3.865	3.836	3.807	3.778	3.749	3.720	3.691
1	4.267	4.235	4.203	4.171	4.140	4.109	4.078	4.047	4.016	3.985
0	4.600	4.565	4.531	4.497	4.463	4.430	4.397	4.364	4.331	4.299
+0	4.600	4.633	4.667	4.700	4.733	4.767	4.801	4.836	4.871	4.905
1	4.940	4.975	5.011	5.047	5.082	5.118	5.155	5.191	5.228	5.265
2	5.302	5.340	5.378	5.416	5.454	5.491	5.530	5.569	5.608	5.647
3	5.687	5.727	5.767	5.807	5.848	5.889	5.930	5.972	6.014	6.055
4	6.097	6.140	6.183	6.226	6.270	6.313	6.357	6.401	6.445	6.490
5	6.534	6.580	6.625	6.671	6.717	6.763	6.810	6.857	6.904	6.951

6	6.998	7.047	7.095	7.144	7.193	7.242	7.292	7.342	7.392	7.443
7	7.492	7.544	7.595	7.647	7.699	7.751	7.804	7.857	7.910	7.964
8	8.017	8.072	8.126	8.181	8.236	8.291	8.347	8.404	8.461	8.517
9	8.574	8.632	8.690	8.748	8.807	8.865	8.925	8.985	9.045	9.105
10	9.165	9.227	9.288	9.350	9.412	9.474	9.537	9.601	9.665	9.728
11	9.792	9.857	9.923	9.989	10.054	10.120	10.187	10.255	10.322	10.389
12	10.457	10.526	10.596	10.665	10.734	10.804	10.875	10.947	11.019	11.090
13	11.162	11.235	11.309	11.383	11.456	11.530	11.605	11.681	11.757	11.832
14	11.906	11.988	12.064	12.142	12.220	12.298	12.378	12.458	12.538	12.619
15	12.699	12.781	12.864	12.947	13.029	13.112	13.197	13.281	13.366	13.451
16	13.535	13.623	13.710	13.797	13.885	13.972	14.062	14.151	14.241	14.331
17	14.421	14.513	14.605	14.697	14.790	14.882	14.977	15.072	15.167	15.262
18	15.360	15.454	15.552	15.650	15.747	15.845	15.945	16.045	16.145	16.246
19	16.346	16.449	16.552	16.655	16.758	16.861	16.967	17.073	17.179	17.285
20	17.391	17.500	17.608	17.717	17.826	17.935	18.047	18.159	18.271	18.383
21	18.495	18.610	18.724	18.839	18.954	19.069	19.187	19.305	19.423	19.541
22	19.650	19.780	19.901	20.022	20.143	20.265	20.389	20.514	20.639	20.763
23	20.888	21.016	21.144	21.272	21.400	21.528	21.659	21.790	21.921	22.053
24	22.184	22.319	22.453	22.588	22.723	22.858	22.996	23.135	23.273	23.411
25	23.550	23.692	23.834	23.976	24.119	24.261	24.406	24.552	24.697	24.842
26	24.988	25.138	25.288	25.438	25.588	25.738	25.891	26.045	26.198	26.351
27	26.505	26.663	26.820	26.978	27.136	27.294	27.455	27.617	27.778	27.939
28	28.101	28.267	28.433	28.599	28.765	28.931	29.101	29.271	29.441	29.612
29	29.782	29.956	30.131	30.305	30.479	30.654	30.833	31.011	31.190	31.369
30	31.548	31.729	31.911	32.094	32.278	32.463	32.650	32.837	33.026	33.215
31	33.405	33.596	33.787	33.980	34.174	34.368	34.564	34.761	34.959	35.159
32	35.359	35.559	35.760	35.962	36.165	36.370	36.576	36.783	36.991	37.200
33	37.410	37.621	37.832	38.045	38.258	38.473	38.689	38.906	39.124	39.344
34	39.565	39.786	40.007	40.230	40.455	40.680	40.907	41.135	41.364	41.595
35	41.827									

TABLE

OF THE TENSION OF THE VAPOR OF MERCURY FOR EACH 10 DEGREES.

REGNAULT. — Mémoires de l'Académie de France, Tome XXVI. p. 520.

Temperature of the Air Thermometer.	Tension.	Temperature of the Air Thermometer.	Tension.	Temperature of the Air Thermometer.	Tension.
°	mm.	°	mm.	°	mm.
0	0.0200	180	11.00	360	797.74
10	0.0268	190	14.84	370	954.65
20	0.0372	200	19.90	380	1139.65
30	0.0530	210	26.35	390	1346.71
40	0.0767	220	34.70	400	1587.96
50	0.1120	230	45.35	410	1863.73
60	0.1643	240	58.82	420	2177.53
70	0.2410	250	75.75	430	2533.01
80	0.3528	260	96.73	440	2933.99
90	0.5142	270	123.01	450	3384.35
100	0.7455	280	155.17	460	3888.14
110	1.0734	290	194.46	470	4449.45
120	1.5341	300	242.15	480	5072.43
130	2.1752	310	299.69	490	5761.32
140	3.0592	320	368.73	500	6520.25
150	4.2664	330	450.91	510	7253.44
160	5.9002	340	548.35	520	8264.96
170	8.0912	350	663.18		

TABLE

FOR DIRECT NITROGEN DETERMINATIONS GIVING $\frac{0.0012562}{(1 + 0.00367\,t)\,760}$.

JAMES T. BROWN. — Jour. Chem. Soc. N. S. Vol. III. p. 210.

t	$\frac{0.0012562}{(1 + 0.00367\,t)\,760}$.	Diff.	t	$\frac{0.0012562}{(1 + 0.00367\,t)\,760}$.	Diff.
0	0.00000165289		16	0.00000156121	544
1	0.00000164685	604	17	0.00000155582	539
2	0.00000164085	600	18	0.00000155047	535
3	0.00000163489	596	19	0.00000154515	532
4	0.00000162898	591	20	0.00000153986	529
5	0.00000162311	587	21	0.00000153462	524
6	0.00000161728	583	22	0.00000152941	521
7	0.00000161149	579	23	0.00000152423	518
8	0.00000160574	575	24	0.00000151909	514
9	0.00000160004	570	25	0.00000151398	511
10	0.00000159438	566	26	0.00000150891	507
11	0.00000158875	563	27	0.00000150387	504
12	0.00000158317	558	28	0.00000149887	500
13	0.00000157762	555	29	0.00000149389	498
14	0.00000157211	551	30	0.00000148896	493
15	0.00000156665	546			

Formula for the use of the above table.

$$W = \frac{0.0012562}{(1 + 0.00367\,t)\,760} \times V \times P.$$

W = weight of gas in grammes.

t = temperature of gas.

V = volume in cc.

P = pressure in mm.

TABLE

OF $\dfrac{0.0012932}{760\,(1 + 0.00367\,t)}$.

JAMES T. BROWN. — Jour. Chem. Soc. N. S. IV. p. 74.

t	A	Diff.	t	A	Diff.
—20.0	0.00000183636	363	2.5	0.00000171733	317
19.5	0.00000183273	361	2.0	0.00000171416	317
19.0	0.00000182912	360	1.5	0.00000171099	315
18.5	0.00000182552	359	1.0	0.00000170784	314
18.0	0.00000182193	357	— 0.5	0.00000170470	313
17.5	0.00000181836	356	0.0	0.00000170157	311
17.0	0.00000181480	355	+ 0.5	0.00000169846	311
16.5	0.00000181125	353	1.0	0.00000169535	309
16.0	0.00000180772	351	1.5	0.00000169226	308
15.5	0.00000180421	351	2.0	0.00000168918	308
15.0	0.00000180070	349	2.5	0.00000168610	306
14.5	0.00000179721	347	3.0	0.00000168304	305
14.0	0.00000179374	347	3.5	0.00000167999	303
13.5	0.00000179027	345	4.0	0.00000167696	303
13.0	0.00000178682	343	4.5	0.00000167393	302
12.5	0.00000178339	343	5.0	0.00000167091	300
12.0	0.00000177996	341	5.5	0.00000166791	300
11.5	0.00000177655	339	6.0	0.00000166491	298
11.0	0.00000177316	339	6.5	0.00000166193	297
10.5	0.00000176977	337	7.0	0.00000165896	297
10.0	0.00000176640	336	7.5	0.00000165599	295
9.5	0.00000176304	334	8.0	0.00000165304	294
9.0	0.00000175970	334	8.5	0.00000165010	293
8.5	0.00000175636	332	9.0	0.00000164717	292
8.0	0.00000175304	330	9.5	0.00000164425	291
7.5	0.00000174974	330	10.0	0.00000164134	290
7.0	0.00000174644	328	10.5	0.00000163844	289
6.5	0.00000174316	327	11.0	0.00000163555	288
6.0	0.00000173989	326	11.5	0.00000163267	287
5.5	0.00000173663	325	12.0	0.00000162980	286
5.0	0.00000173338	323	12.5	0.00000162694	285
4.5	0.00000173015	322	13.0	0.00000162409	284
4.0	0.00000172693	321	13.5	0.00000162125	283
3.5	0.00000172372	320	14.0	0.00000161842	282
3.0	0.00000172052	319	14.5	0.00000161560	281

t	A	Diff.	t	A	Diff.
+15.0	0.00000161279	280	+39.0	0.00000148852	238
15.5	0.00000160999	279	39.5	0.00000148614	238
16.0	0.00000160720	278	40.0	0.00000148376	237
16.5	0.00000160442	277	40.5	0.00000148139	237
17.0	0.00000160165	276	41.0	0.00000147902	235
17.5	0.00000159889	276	41.5	0.00000147667	235
18.0	0.00000159613	274	42.0	0.00000147432	234
18.5	0.00000159339	273	42.5	0.00000147198	233
19.0	0.00000159066	273	43.0	0.00000146965	233
19.5	0.00000158793	271	43.5	0.00000146732	232
20.0	0.00000158522	271	44.0	0.00000146500	231
20.5	0.00000158251	269	44.5	0.00000146269	230
21.0	0.00000157982	269	45.0	0.00000146039	230
21.5	0.00000157713	268	45.5	0.00000145809	229
22.0	0.00000157445	267	46.0	0.00000145580	228
22.5	0.00000157178	266	46.5	0.00000145352	227
23.0	0.00000156912	265	47.0	0.00000145125	227
23.5	0.00000156647	264	47.5	0.00000144898	226
24.0	0.00000156383	263	48.0	0.00000144672	225
24.5	0.00000156120	263	48.5	0.00000144447	225
25.0	0.00000155857	261	49.0	0.00000144222	224
25.5	0.00000155596	261	49.5	0.00000143998	223
26.0	0.00000155335	260	50.0	0.00000143775	445
26.5	0.00000155075	258	51.0	0.00000143330	442
27.0	0.00000154817	258	52.0	0.00000142888	439
27.5	0.00000154559	258	53.0	0.00000142449	436
28.0	0.00000154301	256	54.0	0.00000142013	434
28.5	0.00000154045	255	55.0	0.00000141579	431
29.0	0.00000153790	255	56.0	0.00000141148	428
29.5	0.00000153535	254	57.0	0.00000140720	426
30.0	0.00000153281	253	58.0	0.00000140294	423
30.5	0.00000153028	252	59.0	0.00000139871	421
31.0	0.00000152776	251	60.0	0.00000139450	418
31.5	0.00000152525	251	61.0	0.00000139032	416
32.0	0.00000152274	249	62.0	0.00000138616	413
32.5	0.00000152025	249	63.0	0.00000138203	410
33.0	0.00000151776	248	64.0	0.00000137793	409
33.5	0.00000151528	247	65.0	0.00000137384	406
34.0	0.00000151281	247	66.0	0.00000136978	403
34.5	0.00000151034	245	67.0	0.00000136575	401
35.0	0.00000150789	245	68.0	0.00000136174	399
35.5	0.00000150544	244	69.0	0.00000135775	396
36.0	0.00000150300	243	70.0	0.00000135379	395
36.5	0.00000150057	243	71.0	0.00000134984	391
37.0	0.00000149814	242	72.0	0.00000134593	390
37.5	0.00000149572	240	73.0	0.00000134203	387
38.0	0.00000149332	241	74.0	0.00000133816	385
38.5	0.00000149091	239	75.0	0.00000133431	383

t	A	Diff.	t	A	Diff.
+76°	0.00000133048	381	+124°	0.00000116940	294
77	0.00000132667	379	125	0.00000116646	293
78	0.00000132288	376	126	0.00000116353	291
79	0.00000131912	374	127	0.00000116062	290
80	0.00000131538	372	128	0.00000115772	288
81	0.00000131166	370	129	0.00000115484	287
82	0.00000130796	368	130	0.00000115197	286
83	0.00000130428	366	131	0.00000114911	284
84	0.00000130062	364	132	0.00000114627	283
85	0.00000129698	362	133	0.00000114344	281
86	0.00000129336	360	134	0.00000114063	280
87	0.00000128976	357	135	0.00000113783	278
88	0.00000128619	356	136	0.00000113505	278
89	0.00000128263	354	137	0.00000113227	275
90	0.00000127909	352	138	0.00000112952	275
91	0.00000127557	350	139	0.00000112677	273
92	0.00000127207	348	140	0.00000112404	272
93	0.00000126859	346	141	0.00000112132	270
94	0.00000126513	344	142	0.00000111862	270
95	0.00000126169	343	143	0.00000111592	268
96	0.00000125826	340	144	0.00000111324	266
97	0.00000125486	339	145	0.00000111058	266
98	0.00000125147	337	146	0.00000110792	264
99	0.00000124810	335	147	0.00000110528	263
100	0.00000124475	333	148	0.00000110265	261
101	0.00000124142	332	149	0.00000110004	261
102	0.00000123810	330	150	0.00000109743	259
103	0.00000123480	328	151	0.00000109484	258
104	0.00000123152	326	152	0.00000109226	256
105	0.00000122826	324	153	0.00000108970	256
106	0.00000122502	323	154	0.00000108714	254
107	0.00000122179	321	155	0.00000108460	253
108	0.00000121858	320	156	0.00000108207	252
109	0.00000121538	318	157	0.00000107955	251
110	0.00000121220	316	158	0.00000107704	250
111	0.00000120904	314	159	0.00000107454	248
112	0.00000120590	313	160	0.00000107206	247
113	0.00000120277	311	161	0.00000106959	247
114	0.00000119966	310	162	0.00000106712	245
115	0.00000119656	308	163	0.00000106467	244
116	0.00000119348	306	164	0.00000106223	242
117	0.00000119042	305	165	0.00000105981	242
118	0.00000118737	303	166	0.00000105739	240
119	0.00000118434	302	167	0.00000105499	240
120	0.00000118132	300	168	0.00000105259	239
121	0.00000117832	299	169	0.00000105020	237
122	0.00000117533	297	170	0.00000104783	236
123	0.00000117236	296	171	0.00000104547	236

t	A	Diff.	t	A	Diff.
+172°	0.00000104311	234	+220°	0.00000094145	191
173	0.00000104077	233	221	0.00000093954	190
174	0.00000103844	232	222	0.00000093764	190
175	0.00000103612	231	223	0.00000093574	188
176	0.00000103381	230	224	0.00000093386	188
177	0.00000103151	229	225	0.00000093198	187
178	0.00000102922	228	226	0.00000093011	186
179	0.00000102694	227	227	0.00000092825	185
180	0.00000102467	226	228	0.00000092640	185
181	0.00000102241	225	229	0.00000092455	184
182	0.00000102016	224	230	0.00000092271	183
183	0.00000101792	223	231	0.00000092088	183
184	0.00000101569	222	232	0.00000091905	182
185	0.00000101347	221	233	0.00000091723	181
186	0.00000101126	220	234	0.00000091542	180
187	0.00000100906	219	235	0.00000091362	180
188	0.00000100687	218	236	0.00000091182	179
189	0.00000100469	217	237	0.00000091003	178
190	0.00000100252	217	238	0 00000090825	178
191	0.00000100035	215	239	0.00000090647	177
192	0.00000099820	215	240	0.00000090470	176
193	0.00000099605	213	241	0.00000090294	175
194	0.00000099392	213	242	0.00000090119	175
195	0.00000099179	211	243	0.00000089944	174
196	0.00000098968	211	244	0.00000089770	174
197	0.00000098757	210	245	0.00000089596	173
198	0.00000098547	209	246	0.00000089423	172
199	0.00000098338	208	247	0.00000089251	171
200	0.00000098130	207	248	0.00000089080	171
201	0.00000097923	207	249	0.00000088909	170
202	0.00000097716	205	250	0.00000088739	170
203	0.00000097511	205	251	0.00000088569	169
204	0.00000097306	204	252	0.00000088400	168
205	0.00000097102	203	253	0.00000088232	167
206	0.00000096899	202	254	0.00000088065	167
207	0.00000096697	201	255	0.00000087898	167
208	0.00000096496	201	256	0.00000087731	165
209	0.00000096295	199	257	0.00000087566	165
210	0.00000096096	199	258	0.00000087401	165
211	0.00000095897	198	259	0.00000087236	164
212	0.00000095699	197	260	0.00000087072	163
213	0.00000095502	196	261	0.00000086909	162
214	0.00000095306	196	262	0.00000086747	162
215	0.00000095110	194	263	0.00000086585	162
216	0.00000094916	194	264	0.00000086423	161
217	0.00000094722	193	265	0.00000086262	160
218	0.00000094529	193	266	0.00000086102	159
219	0.00000094336	191	267	0.00000085943	159

t	A	Diff.	t	A	Diff.
+268	0.00000085784	159	+310	0.00000079598	136
269	0.00000085625	158	311	0.00000079462	136
270	0.00000085467	157	312	0.00000079326	136
271	0.00000085310	157	313	0.00000079190	135
272	0.00000085153	156	314	0.00000079055	134
273	0.00000084997	155	315	0.00000078921	134
274	0.00000084842	155	316	0.00000078787	134
275	0.00000084687	155	317	0.00000078653	133
276	0.00000084532	153	318	0.00000078520	133
277	0.00000084379	154	319	0.00000078387	132
278	0.00000084225	152	320	0.00000078255	132
279	0.00000084073	153	321	0.00000078123	132
280	0.00000083920	151	322	0.00000077991	131
281	0.00000083769	151	323	0.00000077860	130
282	0.00000083618	151	324	0.00000077730	130
283	0.00000083467	150	325	0.00000077600	130
284	0.00000083317	149	326	0.00000077470	129
285	0.00000083168	149	327	0.00000077341	129
286	0.00000083019	149	328	0.00000077212	128
287	0.00000082870	148	329	0.00000077084	128
288	0.00000082722	147	330	0.00000076956	128
289	0.00000082575	147	331	0.00000076828	127
290	0.00000082428	146	332	0.00000076701	127
291	0.00000082282	146	333	0.00000076574	126
292	0.00000082136	145	334	0.00000076448	126
293	0.00000081991	145	335	0.00000076322	125
294	0.00000081846	144	336	0.00000076197	125
295	0.00000081702	144	337	0.00000076072	125
296	0.00000081558	143	338	0.00000075947	124
297	0.00000081415	143	339	0.00000075823	124
298	0.00000081272	142	340	0.00000075699	123
299	0.00000081130	141	341	0.00000075576	123
300	0.00000080989	142	342	0.00000075453	123
301	0.00000080847	140	343	0.00000075330	122
302	0.00000080707	141	344	0.00000075208	122
303	0.00000080566	139	345	0.00000075086	121
304	0.00000080427	140	346	0.00000074965	121
305	0.00000080287	138	347	0.00000074844	121
306	0.00000080149	139	348	0.00000074723	120
307	0.00000080010	138	349	0.00000074603	120
308	0.00000079872	137	350	0.00000074483	
309	0.00000079735	137			

TABLE

OF $\dfrac{0.00367}{1 + 0.00367\,t}$.

JAMES T. BROWN. — Jour. Chem. Soc. N. S. IV. p. 74.

t	$\dfrac{0.00367}{1 + 0.00367\,t}$.	Diff.	t	$\dfrac{0.00367}{1 + 0.00367\,t}$.	Diff.
—20.0	0.00396071	783	— 2.5	0.00370398	685
19.5	0.00395288	779	2.0	0.00369713	682
19.0	0.00394509	777	1.5	0.00369031	680
18.5	0.00393732	774	1.0	0.00368351	677
18.0	0.00392958	770	0.5	0.00367674	674
17.5	0.00392188	768	0.0	0.00367000	673
17.0	0.00391420	764	+ 0.5	0.00366327	669
16.5	0.00390656	762	1.0	0.00365658	668
16.0	0.00389894	758	1.5	0.00364990	665
15.5	0.00389136	756	2.0	0.00364325	662
15.0	0.00388380	753	2.5	0.00363663	660
14.5	0.00387627	750	3.0	0.00363003	658
14.0	0.00386877	747	3.5	0 00362345	655
13.5	0.00386130	744	4.0	0.00361690	653
13.0	0.00385386	741	4.5	0.00361037	651
12.5	0.00384645	738	5.0	0.00360386	648
12.0	0.00383907	736	5.5	0.00359738	646
11.5	0.00383171	732	6.0	0.00359092	643
11.0	0.00382439	730	6.5	0.00358449	642
10.5	0 00381709	727	7.0	0.00357807	639
10.0	0.00380982	725	7.5	0.00357168	636
9.5	0.00380257	721	8.0	0 00356532	635
9.0	0.00379536	719	8.5	0.00355897	632
8.5	0.00378817	716	9.0	0 00355265	630
8.0	0.00378101	714	9.5	0.00354635	628
7.5	0.00377387	711	10.0	0.00354007	625
7.0	0.00376676	708	10.5	0.00353382	623
6.5	0.00375968	705	11.0	0.00352759	621
6.0	0.00375263	703	11.5	0 00352138	619
5.5	0.00374560	700	12.0	0.00351519	617
5.0	0.00373860	698	12.5	0.00350902	615
4.5	0.00373162	695	13.0	0.00350287	612
4.0	0.00372467	692	13.5	0.00349675	610
3.5	0.00371775	690	14.0	0.00349065	609
3.0	0.00371085	687	14.5	0.00348456	606

t	$\dfrac{0.00367}{1+0.00367\,t}$	Diff.	t	$\dfrac{0.00367}{1+0.00367\,t}$	Diff.
$+15.0°$	0.00347850	604	$+33.0°$	0.00327354	535
15.5	0.00347246	601	33.5	0.00326819	533
16.0	0.00346645	600	34.0	0.00326286	532
16.5	0.00346045	598	34.5	0.00325754	530
17.0	0.00345447	596	35.0	0.00325224	528
17.5	0.00344851	593	35.5	0.00324696	526
18.0	0.00344258	592	36.0	0.00324170	524
18.5	0.00343666	589	36.5	0.00323646	523
19.0	0.00343077	588	37.0	0.00323123	522
19.5	0.00342489	585	37.5	0.00322601	519
20.0	0.00341904	584	38.0	0.00322082	518
20.5	0.00341320	581	38.5	0.00321564	516
21.0	0.00340739	580	39.0	0.00321048	515
21.5	0.00340159	577	39.5	0.00320533	513
22.0	0.00339582	577	40.0	0.00320020	511
22.5	0.00339005	573	40.5	0.00319509	509
23.0	0.00338432	571	41.0	0.00319000	508
23.5	0.00337861	570	41.5	0.00318492	507
24.0	0.00337291	568	42.0	0.00317985	505
24.5	0.00336723	566	42.5	0.00317480	503
25.0	0.00336157	564	43.0	0.00316977	501
25.5	0.00335593	562	43.5	0.00316476	500
26.0	0.00335031	560	44.0	0.00315976	499
26.5	0.00334471	559	44.5	0.00315477	497
27.0	0.00333912	556	45.0	0.00314980	495
27.5	0.00333356	555	45.5	0.00314485	494
28.0	0.00332801	553	46.0	0.00313991	492
28.5	0.00332248	551	46.5	0.00313499	490
29.0	0.00331697	549	47.0	0.00313009	490
29.5	0.00331148	548	47.5	0.00312519	487
30.0	0.00330600	545	48.0	0.00312032	486
30.5	0.00330055	544	48.5	0.00311546	485
31.0	0.00329511	542	49.0	0.00311061	483
31.5	0.00328969	540	49.5	0.00310578	481
32.0	0.00328429	539	50.0	0.00310097	
32.5	0.00327890	536			

EXAMPLE

		mm.
Observed height Barometer with glass scale		740.0
Attached Thermometer		14.5

	°	mm.
Correction for	10	1.266140
" "	4	0.506460
" "	0.5	0.063307
	Total,	1.835907

_rue height of Barometer   740 — 1.84 = 738.16.

If the temperature is below 0° C. the correction becomes Additive.

The table for a brass scale is used in the same manner.

TABLE

FOR THE REDUCTION OF A BAROMETER WITH A GLASS SCALE TO 0° C.

BUNSEN. — Gasometrischen Tafeln, p. 14.

Millime-tres.	Degrees.									
	1	2	3	4	5	6	7	8	9	10
	mm.	mm.	mm.	mm.	mm.	mm.	mm.	mm.	mm.	mm.
5	0.00085	0.00171	0.00256	0.00342	0.00428	0.00513	0.00599	0.00684	0.00770	0.00855
10	0.00171	0.00342	0.00513	0.00684	0.00856	0.01027	0.01198	0.01369	0.01540	0.01711
15	0.00256	0.00513	0.00769	0.01026	0.01284	0.01540	0.01797	0.02053	0.02310	0.02566
20	0.00342	0.00684	0.01027	0.01368	0.01712	0.02054	0.02396	0.02738	0.03080	0.03422
25	0.00427	0.00855	0.01283	0.01710	0.02139	0.02567	0.02995	0.03422	0.03850	0.04277
30	0.00513	0.01026	0.01540	0.02052	0.02567	0.03080	0.03593	0.04107	0.04620	0.05133
35	0.00598	0.01197	0.01797	0.02394	0.02995	0.03594	0.04192	0.04791	0.05390	0.05988
40	0.00684	0.01368	0.02053	0.02736	0.03423	0.04107	0.04791	0.05476	0.06160	0.06844
45	0.00769	0.01539	0.02310	0.03078	0.03850	0.04620	0.05390	0.06160	0.06930	0.07699
50	0.00855	0.01711	0.02567	0.03422	0.04278	0.05133	0.05989	0.06844	0.07700	0.08555
55	0.00940	0.01882	0.02824	0.03764	0.04706	0.05646	0.06588	0.07528	0.08470	0.09410
60	0.01026	0.02053	0.03080	0.04106	0.05134	0.06160	0.07187	0.08213	0.09240	0.10266
65	0.01111	0.02224	0.03337	0.04449	0.05562	0.06673	0.07786	0.08897	0.10010	0.11121
70	0.01197	0.02395	0.03593	0.04791	0.05990	0.07187	0.08385	0.09582	0.10780	0.11977
75	0.01282	0.02567	0.03850	0.05153	0.06417	0.07700	0.08983	0.10266	0.11550	0.12832
80	0.01368	0.02738	0.04106	0.05476	0.06845	0.08213	0.09582	0.10951	0.12320	0.13688
85	0.01453	0.02909	0.04363	0.05818	0.07273	0.08727	0.10181	0.11635	0.13090	0.14543
90	0.01539	0.03080	0.04619	0.06160	0.07701	0.09240	0.10780	0.12320	0.13860	0.15399
95	0.01625	0.03251	0.04876	0.06502	0.08129	0.09753	0.11378	0.13004	0.14630	0.16254
100	0.01711	0.03422	0.05133	0.06844	0.08555	0.10266	0.11977	0.13688	0.15399	0.17110

105	0.01796	0.03593	0.05390	0.07186	0.08983	0.10779	0.12576	0.14372	0.16169	0.17965
110	0.01882	0.03764	0.05646	0.07528	0.09411	0.11293	0.13175	0.15057	0.16939	0.18821
115	0.01967	0.03935	0.05903	0.07871	0.09839	0.11806	0.13774	0.15741	0.17709	0.19676
120	0.02053	0.04106	0.06160	0.08213	0.10266	0.12320	0.14372	0.16426	0.18479	0.20532
125	0.02138	0.04278	0.06416	0.08555	0.10694	0.12833	0.14971	0.17110	0.19248	0.21387
130	0.02224	0.04449	0.06673	0.08898	0.11122	0.13346	0.15570	0.17795	0.20018	0.22243
135	0.02309	0.04620	0.06929	0.09240	0.11549	0.13860	0.16169	0.18479	0.20788	0.23098
140	0.02395	0.04791	0.07186	0.09582	0.11977	0.14374	0.16767	0.19164	0.21558	0.23954
145	0.02480	0.04962	0.07442	0.09924	0.12405	0.14887	0.17366	0.19848	0.22328	0.24809
150	0.02566	0.05133	0.07699	0.10266	0.12832	0.15399	0.17965	0.20532	0.23098	0.25665
155	0.02651	0.05304	0.07956	0.10608	0.13260	0.15912	0.18564	0.21216	0.23868	0.26520
160	0.02737	0.05475	0.08212	0.10950	0.13688	0.16426	0.19163	0.21901	0.24638	0.27376
165	0.02822	0.05646	0.08469	0.11293	0.14116	0.16939	0.19762	0.22585	0.25408	0.28231
170	0.02908	0.05817	0.08726	0.11635	0.14543	0.17453	0.20361	0.23270	0.26178	0.29087
175	0.02994	0.05989	0.08982	0.11977	0.14971	0.17966	0.20959	0.23954	0.26948	0.29942
180	0.03079	0.06160	0.09239	0.12320	0.15399	0.18479	0.21553	0.24639	0.27718	0.30798
185	0.03165	0.06331	0.09495	0.12662	0.15827	0.18993	0.22157	0.25323	0.28488	0.31653
190	0.03250	0.06502	0.09752	0.13004	0.16254	0.19506	0.22756	0.26008	0.29258	0.32509
195	0.03336	0.06673	0.10008	0.13346	0.16682	0.20019	0.23355	0.26692	0.30028	0.33364
200	0.03422	0.06844	0.10266	0.13688	0.17110	0.20532	0.23954	0.27376	0.30798	0.34220
205	0.03507	0.07015	0.10523	0.14030	0.17538	0.21045	0.24553	0.28060	0.31568	0.35075
210	0.03593	0.07186	0.10779	0.14372	0.17966	0.21559	0.25152	0.28745	0.32338	0.35931
215	0.03678	0.07357	0.11036	0.14715	0.18394	0.22072	0.25751	0.29429	0.33108	0.36786
220	0.03764	0.07528	0.11293	0.15057	0.18821	0.22586	0.26349	0.30114	0.33878	0.37642
225	0.03849	0.07700	0.11549	0.15399	0.19249	0.23099	0.26948	0.30798	0.34647	0.38497
230	0.03935	0.07871	0.11805	0.15742	0.19677	0.23612	0.27547	0.31483	0.35417	0.39353
235	0.04020	0.08042	0.12062	0.16084	0.20105	0.24126	0.28145	0.32167	0.36187	0.40208
240	0.04106	0.08213	0.12318	0.16426	0.20532	0.24639	0.28744	0.32852	0.36957	0.41064
245	0.04191	0.08384	0.12575	0.16768	0.20960	0.25152	0.29343	0.33536	0.37727	0.41919
250	0.04277	0.08555	0.12832	0.17110	0.21387	0.25665	0.29942	0.34220	0.38497	0.42775

Millimetres.	Degrees.									
	1	2	3	4	5	6	7	8	9	10
	mm.	mm.	mm.	mm.	mm.	mm.	mm.	mm.	mm.	mm.
255	0.04362	0.08726	0.13089	0.17452	0.21815	0.26178	0.30541	0.34904	0.39267	0.43630
260	0.04448	0.08897	0.13345	0.17794	0.22243	0.26692	0.31140	0.35589	0.40037	0.44486
265	0.04534	0.09068	0.13601	0.18137	0.22671	0.27205	0.31739	0.36273	0.40807	0.45341
270	0.04619	0.09239	0.13858	0.18479	0.23098	0.27719	0.32338	0.36958	0.41577	0.46197
275	0.04705	0.09411	0.14115	0.18821	0.23526	0.28232	0.32936	0.37642	0.42347	0.47052
280	0.04790	0.09582	0.14371	0.19164	0.23954	0.28745	0.33535	0.38327	0.43117	0.47908
285	0.04876	0.09753	0.14628	0.19306	0.24381	0.29259	0.34134	0.39011	0.43887	0.48763
290	0.04961	0.09924	0.14884	0.19848	0.24809	0.29772	0.34733	0.39696	0.44657	0.49619
295	0.05047	0.10095	0.15141	0.20190	0.25237	0.30285	0.35332	0.40380	0.45427	0.50474
300	0.05133	0.10266	0.15399	0.20532	0.25665	0.30798	0.35931	0.41064	0.46197	0.51330
305	0.05218	0.10437	0.15656	0.20874	0.26093	0.31311	0.36530	0.41748	0.46967	0.52185
310	0.05304	0.10608	0.15912	0.21216	0.26521	0.31825	0.37129	0.42433	0.47737	0.53041
315	0.05389	0.10779	0.16169	0.21559	0.26949	0.32338	0.37728	0.43117	0.48507	0.53896
320	0.05475	0.10951	0.16426	0.21901	0.27376	0.32852	0.38326	0.43802	0.49277	0.54752
325	0.05560	0.11122	0.16682	0.22243	0.27804	0.33365	0.38925	0.44486	0.50046	0.55607
330	0.05646	0.11293	0.16939	0.22586	0.28232	0.33878	0.39524	0.45171	0.50816	0.56463
335	0.05731	0.11464	0.17195	0.22928	0.28659	0.34392	0.40122	0.45855	0.51586	0.57318
340	0.05817	0.11635	0.17452	0.23270	0.29087	0.34905	0.40721	0.46540	0.52356	0.58174
345	0.05902	0.11806	0.17708	0.23612	0.29515	0.35418	0.41320	0.47224	0.53126	0.59029
350	0.05988	0.11977	0.17965	0.23954	0.29942	0.35931	0.41919	0.47908	0.53896	0.59885
355	0.06074	0.12148	0.18222	0.24296	0.30370	0.36444	0.42518	0.48592	0.54666	0.60740
360	0.06159	0.12319	0.18478	0.24638	0.30798	0.36958	0.43117	0.49277	0.55436	0.61596
365	0.06245	0.12490	0.18735	0.24981	0.31226	0.37471	0.43716	0.49961	0.56206	0.62451
370	0.06330	0.12662	0.18992	0.25323	0.31653	0.37985	0.44315	0.50646	0.56976	0.63307
375	0.06416	0.12833	0.19248	0.25665	0.32081	0.38498	0.44913	0.51330	0.57746	0.64162

380	0.65018	0.58516	0.52015	0.45512	0.39011	0.32509	0.26008	0.19505	0.13001	0.06501
385	0.65873	0.59286	0.52699	0.46111	0.39525	0.32937	0.26350	0.19761	0.13175	0.06587
390	0.66729	0.60056	0.53384	0.46710	0.40038	0.33364	0.26692	0.20018	0.13316	0.06672
395	0.67584	0.60826	0.54068	0.47309	0.40552	0.33792	0.27034	0.20274	0.13517	0.06758
400	0.68440	0.61596	0.54752	0.47908	0.41064	0.34220	0.27376	0.20532	0.13688	0.06841
405	0.69295	0.62366	0.55436	0.48507	0.41577	0.34648	0.27718	0.20789	0.13859	0.06929
410	0.70151	0.63136	0.56121	0.49106	0.42091	0.35076	0.28060	0.21045	0.14030	0.07015
415	0.71006	0.63906	0.56805	0.49705	0.42604	0.35504	0.28403	0.21302	0.14201	0.07100
420	0.71862	0.64676	0.57490	0.50303	0.43118	0.35931	0.28745	0.21559	0.14373	0.07186
425	0.72717	0.65445	0.58174	0.50902	0.43631	0.36359	0.29087	0.21815	0.14544	0.07271
430	0.73573	0.66215	0.58859	0.51501	0.44144	0.36787	0.29430	0.22072	0.14715	0.07357
435	0.74428	0.66985	0.59543	0.52099	0.44658	0.37215	0.29772	0.22328	0.14886	0.07442
440	0.75284	0.67755	0.60228	0.52698	0.45171	0.37642	0.30114	0.22585	0.15057	0.07528
445	0.76139	0.68525	0.60912	0.53297	0.45684	0.38070	0.30456	0.22341	0.15228	0.07613
450	0.76995	0.69295	0.61596	0.53896	0.46197	0.38197	0.30798	0.23098	0.15399	0.07699
455	0.77850	0.70065	0.62280	0.54495	0.46710	0.38925	0.31140	0.23355	0.15570	0.07785
460	0.78706	0.70835	0.62965	0.55094	0.47224	0.39353	0.31482	0.23612	0.15741	0.07870
465	0.79561	0.71605	0.63649	0.55693	0.47737	0.39781	0.31825	0.23868	0.15912	0.07956
470	0.80417	0.72375	0.64334	0.56292	0.48251	0.40208	0.32167	0.24125	0.16084	0.08041
475	0.81272	0.73145	0.65018	0.56890	0.48764	0.40636	0.32509	0.24382	0.16255	0.08127
480	0.82128	0.73915	0.65703	0.57489	0.49277	0.41064	0.32852	0.24638	0.16426	0.08212
485	0.82983	0.74685	0.66387	0.58088	0.49791	0.41491	0.33194	0.24895	0.16597	0.08298
490	0.83839	0.75455	0.67072	0.58687	0.50304	0.41919	0.33536	0.25151	0.16768	0.08383
495	0.84694	0.76225	0.67756	0.59286	0.50817	0.42347	0.33878	0.25408	0.16939	0.08469
500	0.85550	0.76995	0.68440	0.59885	0.51330	0.42775	0.34220	0.25665	0.17110	0.08555
505	0.86405	0.77765	0.69124	0.60484	0.51843	0.43203	0.34562	0.25922	0.17281	0.08640
510	0.87261	0.78535	0.69809	0.61083	0.52357	0.43631	0.34904	0.26178	0.17452	0.08726
515	0.88116	0.79305	0.70193	0.61682	0.52870	0.44059	0.35247	0.26435	0.17623	0.08811
520	0.88972	0.80075	0.71178	0.62280	0.53384	0.44486	0.35589	0.26692	0.17795	0.08897
525	0.89827	0.80844	0.71862	0.62879	0.53897	0.44914	0.35931	0.26948	0.17966	0.08982

Millime-tres.	Degrees.									
	1	2	3	4	5	6	7	8	9	10
	mm.	mm.	mm.	mm.	mm.	mm.	mm.	mm.	mm.	mm.
530	0.09068	0.18137	0.27205	0.36274	0.45342	0.54410	0.63478	0.72547	0.81614	0.90683
535	0.09153	0.18308	0.27461	0.36616	0.45769	0.54924	0.64076	0.73231	0.82384	0.91538
540	0.09239	0.18479	0.27718	0.36958	0.46197	0.55437	0.64675	0.73916	0.83154	0.92394
545	0.09324	0.18650	0.27974	0.37300	0.46625	0.55950	0.65274	0.74600	0.83924	0.93249
550	0.09410	0.18821	0.28231	0.37642	0.47052	0.56463	0.65873	0.75284	0.84694	0.94105
555	0.09496	0.18992	0.28488	0.37984	0.47480	0.56976	0.66472	0.75968	0.85464	0.94960
560	0.09581	0.19163	0.28745	0.38326	0.47908	0.57490	0.67071	0.76653	0.86234	0.95816
565	0.09667	0.19334	0.29001	0.38669	0.48336	0.58003	0.67670	0.77337	0.87004	0.96671
570	0.09752	0.19506	0.29258	0.39011	0.48763	0.58517	0.68269	0.78022	0.87774	0.97527
575	0.09838	0.19677	0.29514	0.39353	0.49191	0.59030	0.68867	0.78706	0.88544	0.98382
580	0.09923	0.19848	0.29771	0.39696	0.49619	0.59543	0.69466	0.79391	0.89314	0.99238
585	0.10009	0.20019	0.30027	0.40038	0.50047	0.60057	0.70065	0.80075	0.90084	1.00093
590	0.10094	0.20190	0.30284	0.40380	0.50474	0.60570	0.70664	0.80760	0.90854	1.00949
595	0.10180	0.20361	0.30540	0.40722	0.50902	0.61083	0.71263	0.81444	0.91624	1.01804
600	0.10266	0.20532	0.30798	0.41064	0.51330	0.61596	0.71862	0.82128	0.92394	1.02660
605	0.10351	0.20703	0.31055	0.41406	0.51758	0.62109	0.72461	0.82812	0.93164	1.03515
610	0.10437	0.20874	0.31311	0.41748	0.52186	0.62623	0.73060	0.83497	0.93934	1.04371
615	0.10522	0.21045	0.31568	0.42091	0.52614	0.63136	0.73659	0.84181	0.94704	1.05226
620	0.10608	0.21217	0.31825	0.42433	0.53041	0.63650	0.74257	0.84866	0.95474	1.06082
625	0.10693	0.21388	0.32081	0.42775	0.53469	0.64163	0.74856	0.85550	0.96243	1.06937
630	0.10779	0.21559	0.32337	0.43118	0.53897	0.64676	0.75455	0.86235	0.97013	1.07793
635	0.10864	0.21730	0.32594	0.43460	0.54324	0.65190	0.76053	0.86919	0.97783	1.08648
640	0.10950	0.21901	0.32850	0.43802	0.54752	0.65703	0.76652	0.87604	0.98553	1.09504
645	0.11035	0.22072	0.33107	0.44144	0.55180	0.66216	0.77251	0.88288	0.99323	1.10359
650	0.11121	0.22243	0.33364	0.44486	0.55607	0.66729	0.77850	0.88972	1.00093	1.11215

655	0.11207	0.22414	0.33621	0.44828	0.56035	0.67242	0.78449	0.89656	1.00863	1.12070
660	0.11292	0.22585	0.33877	0.45170	0.56463	0.67756	0.79048	0.90341	1.01633	1.12926
665	0.11378	0.22756	0.34134	0.45513	0.56891	0.68269	0.79647	0.91025	1.02403	1.13781
670	0.11463	0.22928	0.34391	0.45855	0.57318	0.68783	0.80246	0.91710	1.03173	1.14637
675	0.11549	0.23099	0.34647	0.46197	0.57746	0.69296	0.80844	0.92394	1.03943	1.15492
680	0.11634	0.23270	0.34904	0.46540	0.58174	0.69809	0.81443	0.93079	1.04713	1.16348
685	0.11720	0.23441	0.35160	0.46882	0.58602	0.70323	0.82042	0.93763	1.05483	1.17203
690	0.11805	0.23612	0.35417	0.47224	0.59029	0.70836	0.82641	0.94448	1.06253	1.18059
695	0.11891	0.23783	0.35673	0.47566	0.59457	0.71349	0.83240	0.95132	1.07023	1.18914
700	0.11977	0.23954	0.35931	0.47908	0.59885	0.71862	0.83839	0.95816	1.07793	1.19770
705	0.12062	0.24125	0.36188	0.48250	0.60313	0.72375	0.84438	0.96500	1.08563	1.20625
710	0.12148	0.24296	0.36444	0.48592	0.60741	0.72889	0.85037	0.97185	1.09333	1.21481
715	0.12233	0.24467	0.36701	0.48935	0.61169	0.73102	0.85636	0.97869	1.10103	1.22336
720	0.12319	0.24639	0.36958	0.49277	0.61596	0.73916	0.86234	0.98554	1.10873	1.23192
725	0.12404	0.24810	0.37214	0.49619	0.62024	0.74429	0.86833	0.99238	1.11642	1.24047
730	0.12490	0.24981	0.37471	0.49962	0.62452	0.74942	0.87432	0.99923	1.12412	1.24903
735	0.12575	0.25152	0.37727	0.50304	0.62880	0.75456	0.88030	1.00607	1.13182	1.25758
740	0.12661	0.25323	0.37984	0.50646	0.63307	0.75969	0.88629	1.01292	1.13952	1.26614
745	0.12746	0.25494	0.38240	0.50988	0.63735	0.76482	0.89228	1.01976	1.14722	1.27469
750	0.12832	0.25665	0.38497	0.51330	0.64162	0.76995	0.89827	1.02660	1.15492	1.28325
755	0.12918	0.25836	0.38754	0.51672	0.64590	0.77508	0.90426	1.03344	1.16262	1.29180
760	0.13003	0.26007	0.39011	0.52014	0.65018	0.78022	0.91025	1.04029	1.17032	1.30036
765	0.13089	0.26178	0.39267	0.52357	0.65446	0.78535	0.91624	1.04713	1.17802	1.30891
770	0.13174	0.26350	0.39524	0.52699	0.65873	0.79049	0.92223	1.05398	1.18572	1.31747
775	0.13260	0.26521	0.39781	0.53041	0.66301	0.79562	0.92821	1.06082	1.19342	1.32602
780	0.13345	0.26692	0.40037	0.53383	0.66729	0.80075	0.93120	1.06767	1.20112	1.33458
785	0.13431	0.26863	0.40294	0.53725	0.67157	0.80589	0.94019	1.07451	1.20882	1.34313
790	0.13516	0.27034	0.40550	0.54067	0.67584	0.81102	0.94618	1.08136	1.21652	1.35169
795	0.13602	0.27205	0.40707	0.54409	0.68012	0.81615	0.95217	1.08820	1.22422	1.36024
800	0.13688	0.27376	0.41064	0.54752	0.68440	0.82128	0.95816	1.09504	1.23192	1.36880

Degrees.

Millimetres.	1	2	3	4	5	6	7	8	9	10
	mm.	mm.	mm.	mm.	mm.	mm.	mm.	mm.	mm.	mm.
805	0.13773	0.27547	0.41321	0.55094	0.68868	0.82641	0.96415	1.10118	1.23962	1.37735
810	0.13859	0.27718	0.41577	0.55436	0.69296	0.83155	0.97014	1.10873	1.24732	1.38591
815	0.13944	0.27889	0.41834	0.55779	0.69724	0.83668	0.97613	1.11557	1.25502	1.39446
820	0.14030	0.28061	0.42091	0.56121	0.70151	0.84182	0.98211	1.12242	1.26272	1.40302
825	0.14115	0.28232	0.42347	0.56463	0.70579	0.84695	0.98810	1.12926	1.27041	1.41157
830	0.14201	0.28403	0.42604	0.56806	0.71007	0.85208	0.99409	1.13611	1.27811	1.42013
835	0.14286	0.28574	0.42860	0.57148	0.71435	0.85722	1.00007	1.14295	1.28581	1.42868
840	0.14372	0.28745	0.43117	0.57490	0.71862	0.86235	1.00606	1.14980	1.29351	1.43724
845	0.14457	0.28916	0.43373	0.57832	0.72290	0.86748	1.01205	1.15664	1.30121	1.44579
850	0.14543	0.29087	0.43630	0.58174	0.72717	0.87261	1.01804	1.16348	1.30891	1.45435
855	0.14629	0.29258	0.43887	0.58516	0.73145	0.87774	1.02403	1.17032	1.31661	1.46290
860	0.14714	0.29429	0.44144	0.58858	0.73573	0.88288	1.03002	1.17717	1.32431	1.47146
865	0.14800	0.29600	0.44400	0.59201	0.74001	0.88801	1.03601	1.18401	1.33201	1.48001
870	0.14885	0.29772	0.44657	0.59543	0.74428	0.89315	1.04200	1.19086	1.33971	1.48857
875	0.14971	0.29943	0.44914	0.59885	0.74856	0.89828	1.04798	1.19770	1.34741	1.49712
880	0.15056	0.30114	0.45170	0.60228	0.75284	0.90341	1.05397	1.20455	1.35511	1.50568
885	0.15142	0.30285	0.45427	0.60570	0.75712	0.90855	1.05996	1.21139	1.36281	1.51423
890	0.15227	0.30456	0.45683	0.60912	0.76139	0.91368	1.06595	1.21824	1.37051	1.52279
895	0.15313	0.30627	0.45940	0.61354	0.76567	0.91881	1.07194	1.22508	1.37821	1.53134
900	0.15399	0.30798	0.46197	0.61596	0.76995	0.92394	1.07793	1.23192	1.38591	1.53990
905	0.15484	0.30969	0.46454	0.61938	0.77423	0.92907	1.08392	1.23876	1.39361	1.54845
910	0.15570	0.31140	0.46710	0.62280	0.77851	0.93421	1.08991	1.24561	1.40131	1.55701
915	0.15655	0.31311	0.46967	0.62623	0.78279	0.93934	1.09490	1.25245	1.40901	1.56556
920	0.15741	0.31483	0.47224	0.62965	0.78706	0.94448	1.10188	1.25930	1.41671	1.57412
925	0.15826	0.31654	0.47480	0.63307	0.79134	0.94961	1.10787	1.26614	1.42440	1.58267

930	0.15912	0.31825	0.47737	0.63650	0.79562	0.95474	1.11386	1.27299	1.43210	1.59123
935	0.15997	0.31996	0.47993	0.63992	0.79989	0.95988	1.11984	1.27983	1.43980	1.59978
940	0.16083	0.32167	0.48250	0.64334	0.80417	0.96501	1.12583	1.28668	1.44750	1.60834
945	0.16168	0.32338	0.48506	0.64676	0.80845	0.97014	1.13182	1.29352	1.45520	1.61689
950	0.16254	0.32509	0.48763	0.65018	0.81273	0.97527	1.13781	1.30036	1.46290	1.62545
955	0.16339	0.32680	0.49020	0.65360	0.81701	0.98010	1.14380	1.30720	1.47060	1.63400
960	0.16425	0.32851	0.49277	0.65708	0.82129	0.98554	1.14979	1.31405	1.47830	1.64256
965	0.16510	0.33022	0.49533	0.66045	0.82557	0.99067	1.15578	1.32089	1.48600	1.65111
970	0.16596	0.33193	0.49790	0.66387	0.82984	0.99581	1.16177	1.32774	1.49370	1.65967
975	0.16681	0.33364	0.50017	0.66729	0.83412	1.00094	1.16775	1.33458	1.50140	1.66822
980	0.16767	0.33535	0.50303	0.67072	0.83840	1.00607	1.17374	1.34143	1.50910	1.67678
985	0.16852	0.33706	0.50560	0.67414	0.84268	1.01121	1.17973	1.34827	1.51680	1.68533
990	0.16938	0.33877	0.50816	0.67756	0.84695	1.01634	1.18572	1.35512	1.52450	1.69389
995	0.17023	0.34048	0.51073	0.68098	0.85123	1.02147	1.19171	1.36196	1.53220	1.70244
1000	0.17110	0.34220	0.51330	0.68440	0.85550	1.02660	1.19770	1.36880	1.53990	1.71100

TABLE

FOR THE REDUCTION OF THE BAROMETRICAL COLUMN, MEASURED
BY A BRASS SCALE EXTENDING FROM THE CISTERN TO THE
TOP, TO 0° CENTIGRADE.

M. T. DELCROS. —(Guyot's Tables.)

Height of the Barometer.	Degrees, Centigrade.								
	1	**2**	**3**	**4**	**5**	**6**	**7**	**8**	**9**
mm.	mm.	mm.	mm.	mm.	mm.	mm.	mm.	mm.	mm.
665	0.1073	0.2146	0.3220	0.4293	0.537	0.644	0.751	0.859	0.966
670	0.1081	0.2163	0.3244	0.4326	0.541	0.649	0.757	0.865	0.973
675	0.1089	0.2179	0.3268	0.4358	0.545	0.654	0.763	0.871	0.980
680	0.1097	0.2195	0.3292	0.4390	0.549	0.658	0.768	0.878	0.988
685	0.1106	0.2211	0.3317	0.4423	0.553	0.663	0.774	0.884	0.995
690	0.1114	0.2227	0.3341	0.4455	0.557	0.668	0.780	0.891	1.002
695	0.1122	0.2243	0.3365	0.4487	0.561	0.673	0.785	0.897	1.010
700	0.1130	0.2260	0.3389	0.4520	0.565	0.678	0.791	0.904	1.017
705	0.1138	0.2276	0.3414	0.4552	0.569	0.683	0.797	0.910	1.024
710	0.1146	0.2292	0.3438	0.4584	0.573	0.688	0.802	0.917	1.031
715	0.1154	0.2308	0.3462	0.4616	0.577	0.691	0.808	0.923	1.039
720	0.1162	0.2324	0.3486	0.4648	0.581	0.697	0.813	0.930	1.046
725	0.1170	0.2340	0.3510	0.4680	0.585	0.702	0.819	0.936	1.053
730	0.1178	0.2356	0.3535	0.4713	0.589	0.707	0.825	0.943	1.060
735	0.1186	0.2372	0.3559	0.4745	0.593	0.712	0.830	0.949	1.068
740	0.1194	0.2389	0.3583	0.4777	0.597	0.717	0.836	0.955	1.075
745	0.1202	0.2405	0.3607	0.4809	0.601	0.721	0.842	0.962	1.082
750	0.1210	0.2421	0.3631	0.4842	0.605	0.726	0.847	0.968	1.089
755	0.1218	0.2437	0.3655	0.4874	0.609	0.731	0.853	0.975	1.097
760	0.1227	0.2453	0.3680	0.4906	0.613	0.736	0.859	0.981	1.104
765	0.1235	0.2469	0.3704	0.4939	0.617	0.741	0.864	0.988	1.111
770	0.1243	0.2486	0.3728	0.4971	0.621	0.746	0.870	0.994	1.118
775	0.1251	0.2502	0.3752	0.5003	0.625	0.750	0.876	1.001	1.126
780	0.1259	0.2518	0.3777	0.5036	0.629	0.755	0.881	1.007	1.133
785	0.1267	0.2534	0.3801	0.5068	0.633	0.760	0.888	1.014	1.140
790	0.1275	0.2550	0.3825	0.5100	0.637	0.765	0.893	1.020	1.148
795	0.1283	0.2566	0.3849	0.5132	0.641	0.770	0.898	1.026	1.155
800	0.1291	0.2582	0.3874	0.5165	0.646	0.775	0.904	1.033	1.162
805	0.1299	0.2598	0.3898	0.5197	0.650	0.780	0.909	1.039	1.169

TABLE

FOR THE REDUCTION OF WATER PRESSURE TO MERCURY PRESSURE.

BUNSEN. — Gasometrische Methoden.

Water Pressure.	Mercury Pressure.	Water Pressure.	Mercury Pressure.	Water Pressure.	Mercury Pressure.	Water Pressure.	Mercury Pressure.
mm.	mm.	mm.	mm.	mm.	mm.	mm.	mm.
1	0.07	29	2.14	56	4.13	83	6.13
2	0.15	30	2.21	57	4.21	84	6.20
3	0.22	31	2.29	58	4.28	85	6.27
4	0.30	32	2.36	59	4.35	86	6.35
5	0.37	33	2.44	60	4.43	87	6.42
6	0.44	34	2.51	61	4.50	88	6.49
7	0.52	35	2.58	62	4.58	89	6.57
8	0.59	36	2.66	63	4.65	90	6.64
9	0.66	37	.2.73	64	4.72	91	6.72
10	0.74	38	2.80	65	4.80	92	6.79
11	0.81	39	2.88	66	4.87	93	6.86
12	0.89	40	2.95	67	4.94	94	6.94
13	0.96	41	3.03	68	5.02	95	7.01
14	1.03	42	3.10	69	5.09	96	7.08
15	1.12	43 ,	3.17	70	5.17	97	7.16
16	1.18	44	3.25	71	5.24	98	7.23
17	1.26	45	3.32	72	5.31	99	7.31
18	1.33	46	3.39	73	5.39	100	7.38
19	1.40	47	3.47	74	5.46	200	14.76
20	1.48	48	3.54	75	5.54	300	22.14
21	1.55	49	3.62	76	5.61	400	29.52
22	1.62	50	3.69	77	5.68	500	36.90
23	1.70	51	3.76	78	5.76	600	44.28
24	1.77	52	3.84	79	5.83	700	51.66
25	1.84	53	3.91	80	5.90	800	59.04
26	1.92	54	3.99	81	5.98	900	66.42
27	1.98	55	4.06	82	6.05	1000	73.80
28	2.07						

CORRECTION FOR THE DEPRESSION OF THE BARO-

M. T. DELCROS. —

Radius of the tube in Millimetres.	Height of the Meniscus in Millimetres.								
	0 1	0.2	0.3	0.4	0.5	0.6	0.7	0.8	0.9
1.0	1.268	2.460	3.516	4.396	5.085				
1.2	0.876	1.715	2.484	3.162	3.728	4.190			
1.4	0.638	1.256	1.836	2.363	2.825	3.218	3.542		
1.6	0.484	0.955	1.404	1.820	2.196	2.528	2.812	3.050	
1.8	0.378	0.747	1.103	1.437	1.746	2.024	2.270	2 483	2.662
2.0	0.302	0.598	0.885	1.158	1.413	1.648	1.859	2.046	2.209
2.2	0.245	0.487	0.723	0.948	1.161	1.360	1.541	1.705	1.851
2.4	0.203	0.403	0.599	0.787	0.966	1.135	1.292	1.436	1.565
2.6	0.170	0.337	0.502	0.661	0.813	0.958	1.093	1.218	1.332
2.8	0.143	0.285	0.425	0.560	0.691	0.815	0.932	1.041	1.142
3.0	0.122	0.243	0.362	0.478	0.591	0.698	0.800	0.896	0.985
3.2	0.105	0.209	0.312	0.412	0.509	0.602	0.691	0.776	0.855
3.4	0.091	0.181	0.269	0.356	0.441	0.523	0.601	0.675	0.745
3.6	0.079	0.157	0.234	0.310	0.384	0.455	0.524	0.590	0.652
3.8	0.069	0 137	0.205	0.271	0.336	0.399	0.459	0.517	0.572
4.0	0.060	0.120	0.180	0.238	0.295	0.350	0.404	0.455	0.504
4.2	0.053	0.106	0.158	0.210	0.260	0.309	0.356	0.402	0.446
4.4	0.047	0.094	0.140	0.185	0.230	0.273	0.315	0.356	0.395
4.6	0.042	0.083	0.124	0.164	0.204	0.242	0.280	0 316	0.351
4.8	0.037	0.074	0.110	0.146	0.181	0.215	0.249	0.281	0.312
5.0	0.033	0.065	0.098	0.130	0.161	0.192	0.221	0.250	0.278
5.2	0.029	0.058	0.087	0.116	0.144	0.171	0.198	0.224	0.248
5.4	0.026	0.052	0.078	0.103	0.128	0.153	0.177	0.200	0.222
5.6	0.023	0.047	0.070	0.092	0.115	0.137	0.158	0.179	0.199
5.8	0.021	0.042	0.062	0.083	0.103	0.122	0.142	0.160	0.178
6.0	0.019	0.037	0.056	0.074	0.092	0.110	0.127	0.144	0.160
6.2	0.017	0.034	0.050	0.067	0.083	0.099	0.114	0.129	0.144
6.4	0.015	0.030	0.045	0.060	0.074	0.089	0.103	0.116	0.130
6.6	0.014	0.027	0.041	0.054	0.067	0.080	0.093	0.105	0.117
6.8	0.012	0.024	0.037	0.049	0.061	0.072	0.084	0.095	0.105
7.0	0.011	0.022	0.033	0.044	0.055	0.065	0.075	0.085	0.095

METRICAL COLUMN DUE TO CAPILLARY ACTION.

(Guyot's Tables.)

Radius of the tube in Millimetres.	Height of the Meniscus in Millimetres.								
	1.0	1.1	1.2	1.3	1.4	1.5	1.6	1.7	1.8
1.0									
1.2									
1.4									
1.6									
1.8									
2.0	2.348								
2.2	1.978	2.087							
2.4	1.680	1.780	1.866						
2.6	1.436	1.528	1.608	1.676					
2.8	1.235	1.318	1.392	1.456	1.511				
3.0	1.068	1.143	1.210	1.270	1.322	1.368			
3.2	0.928	0.995	1.057	1.112	1.161	1.203	1.238		
3.4	0.810	0.871	0.926	0.976	1.021	1.061	1.095		
3.6	0.710	0.764	0.814	0.860	0.901	0.938	0.970		
3.8	0.624	0.673	0 718	0 760	0.797	0.831	0.861	0.887	
4.0	0.551	0.594	0.635	0.673	0.707	0.738	0.766	0.790	
4.2	0.487	0.526	0.563	0.597	0.628	0.657	0.682	0.705	
4.4	0.432	0 467	0.500	0.531	0.559	0.585	0.609	0 630	
4.6	0.384	0.416	0.445	0.473	0.499	0.522	0.544	0.563	
4.8	0.342	0.370	0.397	0.422	0.445	0.467	0.486	0.504	
5.0	0.305	0.330	0.354	0.377	0.398	0.418	0.436	0.452	
5.2	0.272	0.295	0.317	0.337	0.356	0.374	0.390	0.405	0.418
5.4	0.244	0.264	0 284	0.302	0.319	0.336	0.350	0.364	0.376
5.6	0.218	0.237	0.255	0.271	0.287	0.301	0.315	0.327	0.338
5.8	0.196	0.213	0.228	0 243	0.257	0.271	0.283	0.294	0.304
6.0	0.176	0.191	0.205	0.219	0.231	0.243	0.254	0.264	0.273
6.2	0.158	0.172	0.185	0.197	0.208	0.219	0.229	0.238	0.246
6.4	0.142	0.154	0.166	0.177	0 187	0.197	0.206	0.214	0.221
6.6	0.128	0 139	0.150	0.160	0.169	0 178	0.186	0.193	0.200
6.8	0.116	0.126	0.135	0.144	0.153	0.160	0.168	0.174	0.180
7.0	0.105	0.114	0.122	0.130	0.138	0.145	0.152	0.158	0.163

TABLE

OF THE COEFFICIENTS OF ABSORPTION OF VARIOUS GASES.

BUNSEN. — Gasometrischer Tafeln.

°C.	Nitrogen.		Hydrogen.		Oxygen.	
	In Water.	In Alcohol.	In Water.	In Alcohol.	In Water.	In Alcohol.
0	0.02035	0.12634	0.01930	0.06925	0.04114	0.28397
1	0.01981	0.12593	0.01930	0.06910	0.04007	0.28397
2	0.01932	0.12553	0.01930	0.06896	0.03907	0.28397
3	0.01884	0.12514	0.01930	0.06881	0.03810	0.28397
4	0.01838	0.12476	0.01930	0.06867	0.03717	0.28397
5	0.01794	0.12440	0.01930	0.06853	0.03628	0.28397
6	0.01752	0.12405	0.01930	0.06839	0.03544	0.28397
7	0.01713	0.12371	0.01930	0.06826	0.03465	0.28397
8	0.01675	0.12338	0.01930	0.06813	0.03389	0.28397
9	0.01640	0.12306	0.01930	0.06799	0.03317	0 28397
10	0.01607	0.12276	0.01930	0.06786	0.03250	0.28397
11	0.01577	0.12247	0.01930	0.06774	0.03189	0.28397
12	0.01549	0.12219	0.01930	0.06761	0.03133	0.28397
13	0.01523	0.12192	0.01930	0.06749	0.03082	0.28397
14	0.01500	0.12166	0.01930	0.06737	0.03034	0.28397
15	0.01478	0.12142	0.01930	0.06725	0.02989	0.28397
16	0.01458	0.12119	0.01930	0.06713	0.02949	0.28397
17	0.01441	0.12097	0.01930	0.06701	0.02914	0.28397
18	0.01426	0.12076	0.01930	0.06690	0.02884	0 28397
19	0.01413	0.12056	0.01930	0.06679	0.02858	0.28397
20	0.01403	0.12038	0.01930	0.06668	0 02838	0.28397
21		0.12021	0.01930	0.06657		
22		0.12005	0 01930	0.06646		
23		0.11990	0.01930	0 06636		
24		0.11976	0.01930	0.06626		

cc.	Carbonic Acid.		Carbonic Oxide.		Protoxide of Nitrogen.		Deutoxide of Nitrogen in Alcohol.
	In Water.	In Alcohol.	In Water.	In Alcohol.	In Water.	In Alcohol.	
0	1.7967	4.3295	0.03287	0.20443	1.3052	4.1780	0.31606
1	1.7207	4.2368	0.03207	0.20443	1.2605	4.1088	0.31262
2	1.6481	4.1466	0.03131	0.20443	1.2172	4.0409	0.30928
3	1.5787	4.0589	0.03057	0.20443	1.1752	3.9741	0.30604
4	1.5126	3.9736	0.02987	0.20443	1.1346	3.9085	0.30290
5	1.4497	3.8908	0.02920	0.20443	1.0954	3.8442	0.29985
6	1.3901	3.8105	0.02857	0.20443	1.0575	3.7811	0.29690
7	1.3339	3.7327	0.02796	0.20443	1.0210	3.7192	0.29405
8	1.2809	3.6573	0.02739	0.20443	0.9858	3.6585	0.29130
9	1.2311	3.5844	0.02686	0.20443	0.9520	3.5990	0.28865
10	1.1847	3.5140	0.02635	0.20443	0.9196	3.5408	0.28609
11	1.1416	3.4461	0.02588	0.20443	0.8885	3.4838	0.28363
12	1.1018	3.3807	0.02544	0.20443	0.8588	3.4279	0.28127
13	1.0653	3.3178	0.02504	0.20443	0.8304	3.3734	0.27901
14	1.0321	3.2573	0.02466	0.20443	0.8034	3.3200	0.27685
15	1.0020	3.1993	0.02432	0.20443	0.7778	3.2678	0.27478
16	0.9753	3.1438	0.02402	0.20443	0.7535	3.2169	0.27281
17	0.9519	3.0908	0.02374	0.20443	0.7306	3.1672	0.27094
18	0.9318	3.0402	0.02350	0.20443	0.7090	3.1187	0.26917
19	0.9150	2.9921	0.02329	0.20443	0.6888	3.0714	0.26750
20	0.9014	2.9465	0.02312	0.20443	0.6700	3.0253	0.26592
21		2.9034			0.6525	2.9805	0.26444
22		2.8628			0.6364	2.9368	0.26306
23		2.8247			0.6216	2.8944	0.26178
24		2.7890			0.6082	2.8532	0.26060

CO.	Marsh Gas.		Olefiant Gas.		Ethyl in Water.	Methyl in Water.
	In Water.	In Alcohol.	In Water.	In Alcohol.		
0	0.05449	0.52259	0.2563	3.5950	0.03147	0.0871
1	0.05332	0.51973	0.2473	3.5379	0.03040	0.0838
2	0.05217	0.51691	0.2388	3.4823	0.02947	0.0807
3	0.05104	0.51412	0.2306	3.4280	0.02856	0.0777
4	0.04993	0.51135	0.2227	3.3750	0.02770	0.0748
5	0.04885	0.50861	0.2153	3.3234	0.02689	0.0720
6	0.04778	0.50590	0.2082	3.2732	0.02613	0.0693
7	0.04674	0.50322	0.2018	3.2243	0.02541	0.0668
8	0.04571	0.50057	0.1952	3.1768	0.02474	0.0644
9	0.04470	0.49795	0.1893	3.1307	0.02412	0.0621
10	0.04372	0.49535	0.1837	3.0859	0.02355	0.0599
11	0.04275	0.49278	0.1786	3.0425	0.02303	0.0578
12	0.04180	0.49024	0.1737	3.0005	0.02257	0.0559
13	0.04088	0.48773	0.1693	2.9598	0.02216	0.0541
14	0.03997	0.48525	0.1652	2.9205	0.02179	0.0524
15	0.03909	0.48280	0.1615	2.8825	0.02147	0.0508
16	0.03823	0.48037	0.1583	2.8459	0.02121	0.0493
17	0.03739	0.47798	0.1553	2.8107	0.02100	0.0480
18	0.03657	0.47561	0.1528	2.7768	0.02084	0.0468
19	0.03577	0.47327	0.1506	2.7443	0.02073	0.0457
20	0.03499	0.47096	0.1488	2.7131	0.02065	0.0447
21		0.46867		2.6833		
22		0.46642		2.6549		
23		0.46419		2.6279		
24		0.46199		2.6022		

°C.	Sulphuretted Hydrogen.		Sulphurous Acid.		Ammonia in Water.	Atmospheric Air in Water.
	In Water	In Alcohol.	In Water.	In Alcohol.		
0	4.3706	17.891	68.861	328.62	1049.6	0.02471
1	4.2874	17.242	67.003	311.98	1020.8	0.02406
2	4.2053	16.606	65.169	295.97	993.3	0.02345
3	4.1243	15.983	63.360	280.58	967.0	0.02287
4	4.0442	15.373	61.576	265.81	941.9	0.02237
5	3.9652	14.776	59.816	241.67	917.9	0.02179
6	3.8872	14.193	58.080	238.16	895.0	0.02128
7	3.8103	13.623	56.369	225.25	873.1	0.02080
8	3.7345	13.066	54.683	212.98	852.1	0.02034
9	3.6596	12.523	53.021	201.33	832.0	0.01992
10	3.5858	11.992	51.383	190.31	812.8	0.01953
11	3.5132	11.475	49.770	179.91	794.3	0.01916
12	3.4415	10.971	48.182	170.13	776.6	0.01882
13	3.3708	10.480	46.618	160.98	759.6	0.01851
14	3.3012	10.003	45.079	152.45	743.1	0.01822
15	3.2326	9.539	43.564	144.55	727.2	0.01795
16	3.1651	9.088	42.073	137.27	711.8	0.01771
17	3.0986	8.650	40.608	130.61	696.9	0.01750
18	3.0331	8.225	39.165	124.58	682.3	0.01732
19	2.9687	7.814	37.749	119.17	668.0	0.01717
20	2.9053	7.415	36.216	114.48	654.0	0.01704
21	2.8430	7.030	34.986	110.22	640.2	
22	2.7817	6.659	33.910	106.68	626.5	
23	2.7215	6.300	32.847	103.77	613.0	
24	2.6623	5.955	31.800	101.47	599.5	

ESTIMATION OF THE WEIGHTS OF

Name of Gas.	Formulas.	1000 cc.	2000 cc.	3000 cc.
		Grammes.	Grammes.	Grammes.
Chloride of Acetyl	C_4H_4Cl	2.79386	5.58772	8.38158
Ethyl	C_4H_5	2.59349	5.18698	7.78047
Ammonia	H_3N	0.76271	1.52542	2.28813
Antimony	Sb	23.06332	46.12664	69.18996
Antimoniuretted Hydrogen	H_3Sb	5.90026	11.80052	17.70078
Arsenic	As	13.40892	26.81784	40.22676
Arseniuretted Hydrogen	H_3As	3.48665	6.97330	10.45995
Boron	B	1.94876	3.89752	5.84628
Chloride of Boron	BCl_3	5.24735	10.49470	15.74205
Fluoride of Boron	BF_3	3.06166	6.12332	9.18498
Bromine	Br	6.99990	13.99980	20.99970
Hydrobromic Acid	HBr	3.54471	7.08942	10.63413
Chlorine	Cl	3.17344	6.34688	9.52032
Oxychloride of Carbon	CClO	4.42494	8.84988	13.27482
Hydrochloric Acid	HCl	1.63153	3.26306	4.89459
Cyanogen	C_2N	2.32930	4.65860	6.98790
Chloride of Cyanogen	C_2NCl	2.75137	5.50274	8.25411
Ditetryl	C_8H_8	2.50388	5.00776	7.51164
Elayl	C_4H_4	1.25194	2.50388	3.75582
Fluorine	Fl	1.71634	3.43268	5.14902
Hydrofluoric Acid	HF	0.90298	1.80596	2.70894
Marsh gas	C_2H_4	0.71558	1.43116	2.14674
Iodine	I	11.36180	22.72360	34.08540
Hydriodic Acid	HI	5.72573	11.45146	17.17719
Silicon	Si	3.80814	7.61628	11.42442
Fluoride of Silicon	SiF_3	4.70206	9.40412	14.10618
Carbon	C	1.07272	2.14544	3.21816
Carbonic Oxide	CO	1.25150	2.50300	3.75450
Carbonic Acid	CO_2	1.97741	3.99482	5.93223
Methyl	C_2H_3	1.34152	2.68304	4.02456
Oxide of Methyl	C_2H_3O	2.05669	4.11338	6.17007
Chloride of Methyl	C_2H_3Cl	2.25749	4.51498	6.77247
Phosphorus	P	5.54230	11.08460	16.62690
Phosphuretted Hydrogen	H_3P	1.52000	3.04000	4.56000

4000 cc.	5000 cc.	6000 cc.	7000 cc.	8000 cc.	9000 cc.
Grammes.	Grammes.	Grammes.	Grammes.	Grammes.	Grammes.
11.17544	13.96930	16.76316	19.55702	22.35088	25.14474
10.37396	12.96745	15.56094	18.15443	20.74792	23.34141
3.05084	3.81355	4.57626	5.33897	6.10168	6.86439
92.25328	115.31660	138.37992	161.44324	184.50656	207.56992
23.60104	29.50130	35.40156	41.30182	47.20208	53.10234
53.63568	67.04460	80.45352	93.86244	107.27136	120.68028
13.94660	17.43325	20.91990	24.40650	27.89320	31.37985
7.79504	9.74380	11.69256	13.64132	15.59008	17.53884
20.98940	26.23675	31.48410	36.73145	41.97880	47.22615
12.24664	15.30830	18.36996	21.43162	24.49328	27.55494
27.99960	34.99950	41.99940	48.99930	55.99920	62.99910
14.17884	17.72355	21.26826	24.81297	28.35768	31.90239
12.69376	15.86720	19.04064	22.21408	25.38752	28.56096
17.69976	22.12470	26.54964	30.97458	35.39952	39.82446
6.52612	8.15765	9.78918	11.42071	13.05224	14.68377
9.31720	11.64650	13.97580	16.30510	18.63440	20.96370
11.00548	13.75685	16.50822	19.25959	22.01096	24.76233
10.01552	12.51940	15.02328	17.52716	20.03104	22.53492
5.00776	6.25970	7.51164	8.76356	10.01552	11.26746
6.86536	8.58170	10.29804	12.01438	13.73072	15.44706
3.61192	4.51490	5.41788	6.32086	7.22384	8.12682
2.86232	3.57790	4.29348	5.00906	5.72464	6.44022
45.44720	56.80900	68.17080	79.53260	90.89440	102.25620
22.90292	28.62865	34.35438	40.08011	45.80584	51.53157
15.23256	19.04070	22.84884	26.65698	30.46512	34.27326
18.80824	23.51030	28.21236	32.91442	37.61648	42.31854
4.29088	5.36360	6.43632	7.50904	8.58176	9.65448
5.00600	6.25750	7.50900	8.76050	10.01200	11.26350
7.98964	9.88705	1.86476	13.84187	15.81928	17.79669
5.36608	6.70760	8.04912	9.39064	10.73216	12.07368
8.22676	10.28345	12.34014	14.39683	16.45352	18.51021
9.02996	11.28745	13.54494	15.80243	18.05992	20.31741
22.16920	27.71150	33.25380	38.79610	44.33840	49.88070
6.08000	7.60000	9.12000	10.64000	12.16000	13.68000

Name of Gas.	Formulas.	1000 cc.	2000 cc.	3000 cc.
		Grammes.	Grammes.	Grammes.
Oxygen	O	1.42980	2.85960	4.28941
Sulphur	S	17.16336	34.32672	51.49008
Sulphurous Acid	SO_2	· 2.86056	5.72112	8.58168
Sulphuretted Hydrogen	HS	1.51991	3.03982	4.55973
Selenium	Se	7.02556	14.05112	21.07668
Selenuretted Hydrogen	HSe	3.60239	7.20478	10.80717
Nitrogen	N	1.25618	2.51236	3.76854
Deutoxide of Nitrogen	NO_2	1.34343	2.68686	4.03029
Protoxide of Nitrogen	NO	1.97172	3.94344	5.91516
Tellurium	Te	11.53525	23.07050	34.60575
Telluretted Hydrogen	HTe	5.85723	11.71446	17.57169
Water Vapor	HO	0.80475	1.60950	2.41425
Hydrogen	H	0.08958	0.17916	0.26874
Atmospheric Air		1.29319	2.58638	3.87957

ESTIMATION OF THE VOLUME OF OXYGEN

Volume of Air.	100.00.	200.00.	300.00.
Volume of Nitrogen	79.04	158.08	237.12
Volume of Oxygen	20.96	41.92	62.88

4000 cc.	5000 cc.	6000 cc.	7000 cc.	8000 cc.	9000 cc.
Grammes.	Grammes.	Grammes.	Grammes.	Grammes.	Grammes.
1.71921	7.14901	8.57881	10.00861	11.41842	12.86822
68.65344	85.81680	102.98016	120.14352	137.30688	154.47024
11.44224	14.30280	17.16336	20.02392	22.88448	25.74504
6.07964	7.59955	9.11946	10.63937	12.15928	13.67919
28.10224	35.12780	42.15336	49.17892	56.20448	63.23004
14.40956	18.01195	21.61434	25.21673	28.81912	32.42151
5.02572	6.28090	7.53708	8.79326	10.04944	11.30562
5.37372	6.71615	8.06058	9.40301	10.74744	12.09087
7.88688	9.85860	11.83032	13.80204	15.77376	17.74548
46.14100	57.67625	69.21150	80.74675	92.28200	103.81725
23.42892	29.28615	35.14338	41.00061	46.85784	52.71507
3.21900	4.02375	4.82850	5.63325	6.43800	7.24275
0.35832	0.44790	0.53748	0.62706	0.71664	0.80622
5.17276	6.46595	7.75914	9.05233	10.34562	11.63871

AND NITROGEN IN ATMOSPHERIC AIR.

400 00.	500.00.	600.00.	700.00.	800.00.	900.00.
316.16	395.20	474.24	553.28	632.32	711.36
83.84	104.80	125.76	146.72	167.68	188.64

TABLES

RELATING TO LIGHT.

If d represent the density and n the index of refraction of a liquid for a given ray of light, the quantity

$$\frac{n-1}{d}$$

is found to be constant, and is termed the specific refractive power. The relation between the specific refractive power of a mixture and the specific refractive powers of its constituents is given by the equation,

$$\frac{N-1}{D}\cdot P = \frac{n-1}{d}\cdot p + \frac{n'-1}{d'}\cdot p' + \frac{n''-1}{d''}\cdot p'' + \ldots$$

where, p, p', p'', &c. represent the weights of the constituents in P units of weight of the mixture.

Hence, if the specific refractive power of a mixture be determined at any temperature from its index of refraction and density, and if the corresponding values are known for each constituent, the relative weights of the latter may easily be calculated when only two are present.

If P be taken as 100 we have

$$\frac{n-1}{d}\cdot p + \frac{n'-1}{d'}(100-p) = \frac{N-1}{D}\cdot 100,$$

whence,

$$p = \frac{100\left(\frac{N-1}{D}-\frac{n'-1}{d'}\right)}{\frac{n-1}{d}-\frac{n'-1}{d'}},$$

$$p' = 100 - p.$$

TABLE

OF WAVE LENGTHS IN MILLIONTHS OF MILLIMETRES.

Those marked * are Ångström's measurements.

DITSCHEINER. — Sitzungsberichte der Kaiserlichen Akademie der Wissenschaften, Bd. L. Heft III. Jahrgang, 1864. Corrected according to the Wave Length of D α given by Ångström. (Pogg. Ann. CXXIII. s. 489.)

Kirchhoff's Line.	Wave Length.	Kirchhoff's Line.	Wave Length	Kirchhoff's Line.	Wave Length	Kirchhoff's Line.	Wave Length.
A	*761.20	1242.6	557.77	1737.7	511.42	2221.7	474.34
B 592.7	*687.49	1280.0	553.27	1750.4	510.30	2233.7	473.36
C 694.1	*656.77	1303.5	551.13	1777.5	508.41	2250.0	472.26
711.4	652.09	1306.7	550.70	1799.0	506.91	2264.3	470.71
719.6	650.05	1324.8	548.13	1834.3	504.55	2309.0	467.07
783.8	634.05	1337.0	546.78	1854.9	503.25	2416.0	460.63
831.0	623.61	1343.5	546.05	1867.1	502.26	2436.5	458.66
849.7	619.72	1351.1	545.08	1873.4	501.67	2457.5	456.81
860.2	617.50	1367.0	543.42	1885.8	501.07	2467.6	455.71
863.9	616.73	1389.4	540.91	1908.5	499.72	2489.4	453.73
874.3	614.74	1410.5	538.79	1920.2	498.78	2537.1	450.56
877.0	614.24	1421.5	537.52	1960.8	496.15	2547.2	450.13
884.9	612.76	1449.4	534.41	1975.7	495.04	2566.3	448.46
894.9	610.82	1463.3	533.29	1983.3	494.34	2606.6	446.00
959.8	598.14	1492.4	530.22	1989.5	493.76	2627.0	444.65
Dβ 1002.8	*590.04	1506.3	528.79	2001.6	492.22	2638.5	443.85
Dγ 1004.8	*589.74	1515.5	528.02	2018.5	491.41	2670.0	442.29
Dα 1006.8	*589.43	E 1523.7	*527.38	2041.3	489.54	2686.6	440.87
1029.3	586.26	1541.9	525.98	2058.0	488.18	2721.6	438.75
1096.1	576.72	1569.6	523.74	2067.1	487.55	2734.1	437.80
1102.9	575.78	1577.6	523.09	F 2080.0	*486.52	2775.7	435.67
1135.1	571.51	1589.1	521.97	2103.3	484.63	2796.7	434.34
1155.7	568.70	1601.7	521.32	2119.8	482.81	2822.3	432.82
1174.2	566.33	1622.3	519.67	2148.9	480.56	G 2854.7	*431.03
1200.6	562.97	1634.1	518.73	2157.4	479.55	2869.7	430.37
1207.3	561.98	b 1648.8	517.70	2160.6	479.23	H	*397.16
1217.8	560.77	1655.6	517.15	2187.1	476.93	H'	*393.59
1237.8	559.13	1693.8	514.65	2201.9	475.90		

TABLE

OF WAVE LENGTHS OF VARIOUS WELL-MARKED LINES OF THE SPECTRUM IN MILLIONTHS OF A MILLIMETRE.

W. GIBBS. — American Jour. of Science, XXXIX. 217.

Line.	Wave Length.	Observer.	Line.	Wave Length.	Observer.
a L	676.3	Müller	ϵ I	516.7	Plücker
β N	661.0	"	γ SiCl$_2$	505.0	"
a H	653.3	Plücker	β H	484.3	"
a PCl$_3$	649.3	"	β Br	479.3	"
a SnCl$_2$	644.5	"	γ Cl	479.2	"
a SiCl$_2$	632.9	"	γ Br	476.6	"
a O	615.0	"	δ Br	469.1	"
11 N	608.9	"	ζ I	466.1	"
β PCl$_3$	602.4	"	δ S$_2$	463.1	Müller
β SiCl$_2$	597.8	"	η I	462.9	Plücker
β I	594.7	"	γ PCl$_3$	459.1	"
β SnCl$_2$	579.4	"	ϵ SnCl$_2$	452.4	"
a Hg	578.2	"	ζ CO$_2$	450.1	"
17 N	576.2	"	θ I	444.6	"
a' Hg	575.9	"	η CO$_2$	438.2	"
γ CO$_2$	559.9	"	δ O	436.7	"
γ SnCl$_2$	558.4	"	γ Hg	435.9	"
β Hg	546.1	"	γ H	433.9	"
a Cl	545.1	"	ι I	421.5	"
δ I	533.7	"	L	379.1	Esselbach
δ SnCl$_2$	533.3	"	M	365.7	"
β O	532.8	"	N	349.8	"
β Cl	521.6	"	O	336.0	"
δ CO$_2$	519.0	"	P	329.0	"
γ O	518.5	"	Q	323.2	"
a Br	516.9	"	R	309.1	"

INDICES OF

OF SOME OF THE MOST

$\Delta = $ *The diminution in refraction for*

LANDOLT. — Pogg. Ann., Bd.

Compounds.	Formula.		Combining Number. P.	Specific Gravity. d.	μ_a
Water	HO		9	1.0000	1.331110
		Δ			0.000096
Formic Acid	$C_2H_2O_4$		46	1.2211	1.369270
		Δ			0.000395
Acetic Acid	$C_4H_4O_4$		60	1.0514	1.369850
		Δ			0.000418
Propionic Acid	$C_6H_6O_4$		74	0.9963	1.384600
		Δ			0.000399
Butyric Acid	$C_8H_8O_4$		88	0.9610	1.395540
		Δ			0.000412
Valeric Acid	$C_{10}H_{10}O_4$		102	0.9313	1.402200
		Δ			0.000406
Capronic Acid	$C_{12}H_{12}O_4$		116	0.9252	1.411640
		Δ			0.000396
Œnanthic Acid	$C_{14}H_{14}O_4$		130	0.9175	1.419230
		Δ			0.000391
Methyl Alcohol	$C_2H_4O_2$		32	0.7964	1.327890
		Δ			0.000380
Ethyl Alcohol	$C_4H_6O_2$		46	0.8011	1.360540
		Δ			0.000400
Propyl Alcohol	$C_6H_8O_2$		60	0.8042	1.379380
Butyl Alcohol	$C_8H_{10}O_2$		74	0.8074	1.393950
		Δ			0.000390
Amyl Alcohol	$C_{10}H_{12}O_2$		88	0.8135	1.405730
		Δ			0.000390
Acetate of Methyl	$C_6H_6O_4$		74	0.9053	1.359150
		Δ			0.000520
Formate of Ethyl	$C_6H_6O_4$		74	0.9078	1.358000
		Δ			0.000530
Acetate of Ethyl	$C_8H_8O_4$		88	0.9021	1.370680
		Δ			0.000500
Butyrate of Methyl	$C_{10}H_{10}O_4$		102	0.8976	1.386930
		Δ			0.000490

REFRACTION

COMMON ORGANIC COMPOUNDS.

an elevation of temperature of one degree.

117, S. 353, Bd. 122, S. 545.

μ_β	μ_γ	A	B	$\frac{\mu_a - 1}{d}$	$P\left(\frac{\mu_a - 1}{d}\right)$
1.337120	1.340380	1.323920	0.30997	0.3311	5.96
0.000108	0.000110				
1.376430	1.380410	1.360620	0.37250	0.3024	13.91
0.000400	0.000433	0.000368			
1.376480	1.380170	1.361840	0.34588	0.3518	21.11
0.000408	0.000430	0.000409			
1.391290	1.395130	1.376430	0.35210	0.3860	28.57
0.000402	0.000402	0.000397			
1.402460	1.406490	1.387040	0.36614	0.4116	36.22
0.000419	0.000429	0.000398			
1.409310	1.413490	1.393440	0.37751	0.4319	44.05
0.000420	0.000423	0.000393			
1.419000	1.423230	1.402640	0.38754	0.4449	51.61
0.000409	0.000413	0.000380			
1.426630	1.431060	1.410050	0.39557	0.4569	59.40
0.000411	0.000410	0.000375			
1.333200	1.336210	1.321430	0.27821	0.4117	13.17
0.000400	0.000400				
1.366650	1.369970	1.353220	0.31532	0.4501	20.70
0.000410	0.000410				
1.385810	1.389320	1.371670	0.33238	0.4717	28.30
1.400690	1.404470	1.385790	0.35177	0.4879	36.11
0.000410	0.000410				
1.412780	1.416890	1.397070	0.37317	0.4987	43.89
0.000400	0.000420				
1.365390	1.368930	1.351560	0.32702	0.3967	29.36
0.000520	0.000530				
1.364200	1.367820	1.350380	0.32836	0.3944	29.18
0.000550	0.000570				
1.377090	1.380670	1.362930	0.33405	0.4109	36.16
0.000520	0.000540				
1.393590	1.397420	1.378790	0.35077	0.4311	43.97
0.000510	0.000520				

Compounds.	Formula.	Combining Number. $P.$	Specific Gravity. $d.$	μ_a
Valerate of Methyl	$C_{12}H_{12}O_4$	116	0.8809	1.392720
Δ				0.000460
Butyrate of Ethyl	$C_{12}H_{12}O_4$	116	0.8906	1.394040
Δ				0.000480
Formate of Amyl	$C_{12}H_{12}O_4$	116	0.8816	1.395920
Δ				0.000480
Valerate of Ethyl	$C_{14}H_{14}O_4$	130	0.8674	1.395000
Δ				0.000470
Acetate of Amyl	$C_{14}H_{14}O_4$	130	0.8574	1.401680
Δ				0.000430
Valerate of Amyl	$C_{20}H_{20}O_4$	172	0.8581	1.409780
Aldehyde	$C_4H_4O_2$	44	0.7810	1.329750
Δ				0.000580
Valeral	$C_{10}H_{10}O_2$	86	0.7995	1.386140
Δ				0.000470
Aceton	$C_6H_6O_2$	58	0.7931	1.357150
Δ				0.000520
Ether	$C_8H_{10}O_2$	74	0.7166	1.351120
Δ				0.000580
Anhydrous Acetic Acid	$C_8H_6O_6$	102	1.0836	1.388320
Δ				0.000460
Glycol	$C_4H_6O_4$	62	1.1092	1.425300
Δ				0.000280
Binacetate of Glycol	$C_{12}H_{10}O_8$	146	1.1583	1.419320
Glycerin	$C_6H_8O_6$	92	1.2615	1.470630
Δ				0.000200
Lactic Acid	$C_6H_6O_6$	90	1.2427	1.439150
Δ				0.000370
Phenic Acid	$C_{12}H_6O_2$	94	1.0722	1.544470
Δ				0.000420
Bitter Almond Oil	$C_{14}H_6O_2$	106	1.0474	1.539140
Δ				0.000500
Salicylous Acid	$C_{14}H_6O_4$	122	1.1693	1.564670
Δ				0.000490
Methylsalicylic Acid	$C_{16}H_8O_6$	152	1.1824	1.530190
Δ				0.000440
Benzoate of Methyl	$C_{16}H_8O_4$	136	1.0882	1.511580
Δ				0.000450
Benzoate of Ethyl	$C_{18}H_{10}O_4$	150	1.0491	1.501040
Δ				0.000460

μ_a, μ_β, μ_γ = the Indices of Refraction for the three Hydrogen lines, $H\alpha$, $H\beta$, and $H\gamma$.

μ_β	μ_γ	A	B	$\dfrac{\mu_a - 1}{d}$	$P\left(\dfrac{\mu_a - 1}{d}\right)$
1.399690	1.403700	1.384200	0.36715	0.4458	51.71
0.000470	0.000480				
1.400730	1.404600	1.385800	0.35310	0.4424	51.32
0.000490	0.000500				
1.402690	1.406890	1.387410	0.36682	0.4491	52.09
0.000500	0.000510				
1.401870	1.405830	1.386590	0.36214	0.4554	59.20
0.000480	0.000490				
1.408760	1.412710	1.393120	0.36882	0.4685	60.90
0.000430	0.000440				
1.417120	1.421240	1.400890	0.38320	0.4775	82.14
1.335880	1.339370	1.322290	0.32161	0.4222	18.58
0.000610	0.000620				
1.393360	1.397290	1.377490	0.37283	0.4830	41.54
0.000500	0.000520				
1.363920	1.367800	1.348880	0.35612	0.4503	26.12
0.000540	0.000550				
1.357200	1.360710	1.343680	0.32067	0.4900	36.26
0.000590	0.000590				
1.395250	1.399270	1.379820	0.36614	0.3584	36.56
0.000470	0.000490				
1.432510	1.436620	1.416510	0.37832	0.3834	23.77
0.000320	0.000370				
1.426810	1.431200	1.410100	0.39725	0.3620	52.85
1.478450	1.482810	1.461180	0.40728	0.3731	34.32
0.000220	0.000240				
1.446860	1.451350	1.429680	0.40794	0.3534	31.81
0.000380	0.000380				
1.563570	1.575550	1.520350	1.03925	0.5078	47.73
0.000440	0.000470				
1.562350	1.577490	1.509400	1.28201	0.5147	54.56
0.000510	0.000540				
1.596000	1.620080	1.521670	1.85280	0.4829	58.91
0.000520	0.000540				
1.552120	1.567180	1.501480	1.23687	0.4484	68.16
0.000460	0.000510				
1.528900	1.539890	1.489610	0.94663	0.4701	63.94
0.000490	0.000500				
1.517150	1.527490	1.480510	0.88444	0.4776	71.64
0.000510	0.000550				

$A =$ Coefficient of Refraction.

$B =$ Coefficient of Dispersion in Cauchy's formula

$$\mu = A + \frac{B}{\lambda^2}.$$

TABLE

OF INDICES OF REFRACTION OF ESSENTIAL OILS.

GLADSTONE. — Journal of the Chemical Society of London, N. S. Vol. II. p. 1.

Crude Oils.	Specific Gravity 15.5 °C.	Temp.	A	D	H	Rotation.
Anise	0.9852	16.5	1.5433	1.5566	1.6118	− 1.0
Antherosperma Moschatum	1.0425	14.0	1.5172	1.5274	1.5628	+ 7.0
Bay	0.8808	18.5	1.4944	1.5022	1.5420	− 6.0
Bergamot	0.8825	22.0	1.4559	1.4625	1.4779G	+ 23.0
" Florence	0.8804	26.5	1.4547	1.4614	1.4760G	+ 40.0
Birch Bark	0.9005	8.0	1.4851	1.4921	1.5172	+ 38.0
Cajeput	0.9203	25.5	1.4561	1.4611	1.4778	0.0
Calamus	0.9388	10.0	1.4965	1.5031	1.5204G	+ 43.5
" Hamburg	0.9410	11.0	1.4843	1.4911	1.5144	+42.0?
Caraway	0.8845	19.0	1.4601	1.4671	1.4886	+ 63.0
" Hamburg 1st dist.	0.9121	10.0	1.4829	1.4903	1.5142	
" " 2d "	0.8832	10.5		1.4784		
Cascarilla	0.8956	10.0	1.4844	1.4918	1.5158	+ 26.0
Cassia	1.0297	19.5	1.5602	1.5748	1.6243G	0.0
Cedar	0.9622	23.0	1.4978	1.5035	1.5238	+ 3.0
Cedrat	0.8584	18.0	1.4671	1.4731	1.4952	+156.0
Citronella	0.8908	21.0	1.4599	1.4659	1.4866	− 4.0
" Penang	0.8847	15.5	1.4604	1.4665	1.4875	− 1.0
Cloves	1.0475	17.0	1.5213	1.5312	1.5666	− 4.0
Coriander	0.8775	10.0	1.4592	1.4652	1.4805G	+21.0?
Cubebs	0.9414	10.0	1.4953	1.5011	1.5160G	
Dill	0.8922	11.5	1.4764	1.4834	1.5072	+206.0
Elder	0.8584	8.5	1.4686	1.4749	1.4965	+ 14.5
Eucalyptus amygdalina	0.8812	13.5	1.4717	1.4788	1.5021	−136.0
" oleosa	0.9322	13.5	1.4661	1.4718	1.4909	+ 4.0
Indian Geranium	0.9043	21.5	1.4653	1.4714	1.4868G	− 4.0
Lavender	0.8903	20.0	1.4586	1.4648	1.4862	− 20.0
Lemon-grass	0.8932	24.0		1.4705		− 3.0?
" Penang	0.8766	13.5	1.4756	1.4837	1.5042	0.0
Malaleuca ericifolia	0.9030	9.0	1.4655	1.4712	1.4901	+ 26.0
" linarifolia	0.9016	9.0	1.4710	1.4772	1.4971	+ 11.0
Mint	0.9342	19.0	1.4767	1.4840	1.5015G	−116.0
"	0.9105	14.5	1.4756	1.4822	1.5037	− 13.0

Crude Oils.	Specific Gravity 15.5 °C.	Refractive Indices				Rotation.
		Temp.	A	D	H	
Myrtle	0.8911	14.0	1.4623	1.4680	1.4879	+ 21.0
Myrrh	1.0189	7.5	1.5196	1.5278	1.5472G	—136.0
Neroli	0.8789	18.0	1.4614	1.4676	1.4835G	+ 15.0
"	0.8743	10.0	1.4673	1.4741	1.4831F	+ 28.0
Nutmeg	0.8826	24.0	1.4644	1.4709	1.4934	+ 44.0
" Penang	0.9069	16.0	1.4749	1.4818	1.5053	+ 9.0
Orange-peel	0.8509	20.0	1.4633	1.4699	1.4916	+32.0?
" Florence	0.8864	20.0	1.4707	1.4774	1.4980	+216.0
Parsley	0.9926	8.5	1.5068	1.5162	1.5417G	— 9.0
Patchouli	0.9554	21.0	1.4990	1.5050	1.5194G	
" Penang	0.9592	21.0	1.4980	1.5040	1.5183G	—120.0
" French	1.0119	14.0	1.5074	1.5132	1.5202F	
Peppermint	0.9028	14.5	1.4612	1.4670	1.4854	— 72.0
" Florence	0.9116	14.0	1.4628	1.4682	1.4867	— 44.0
Petit Grain	0.8765	21.0	1.4536	1.4600	1.4808	+ 26.0
Rose	0.8912	25.0	1.4567	1.4627	1.4835	— 7.0
Rosemary	0.9080	16.5	1.4632	1.4688	1.4867	+ 17.0
Rosewood	0.9064	17.0	1.4843	1.4903	1.5113	— 16.0
Sandalwood	0.9750	24.0	1.4959	1.5021	1.5227	— 50.0
Thyme	0.8843	19.0	1.4695	1.4754	1.4909G	
Turpentine	0.8727	13.0	1.4672	1.4732	1.4938	— 79.0
Verbena	0.8812	20.0	1.4791	1.4870	1.5059G	— 6.0
Wintergreen	1.1423	15.0	1.5163	1.5278	1.5737	+ 3.0
Wormwood	0.9122	18.0	1.4631	1.4688	1.4756F	

NOTE. — The tube by which the Angle of Rotation was determined was ten inches long. The numbers marked G and F are the indices of refraction for the lines G and F, the color of the liquid preventing H from being seen.

TABLE

OF THE PHYSICAL PROPERTIES OF THE HYDROCARBONS OF THE ESSENTIAL OILS.

GLADSTONE. — Journal of the Chemical Society of London, 2d Series, Vol. II. p 1.

Source of Hydrocarbon.	Sp. Gr. at 20° C. = D	Boiling Point.	Index of Refraction of A at 20° C. $\mu A =$	Dispersion. A to H	$\Delta 10°$	$\frac{\mu A - 1}{D}$	Rotation.
Orange peel	.8460	174°	1.4645	.0277	.0048	.5490	+154°
" " Florence	.8468	174	1.4650	.0281	.0049	.5491	+260
Cedrat	.8466	173	1.4650	.0280	.0049	.5492	+180
Lemon	.8468	173	1.4660	.0280	0049	.5502	+172
Bergamot	.8466	175	1.4619	.0295	.0049	.5456	+ 76
" Florence	.8464	176	1.4602	.0287	.0048	.5437	+ 82
Neroli	.8466	173	1.4614	.0291	.0047	.5450	+ 76
Petit Grain	.8470	174	1.4617	.0282	.0046	.5439	+ 60
Caraway Hamburg 1st dist.	.8466	176	1.4645	.0286	.0048	.5486	+180
Dill	.8467	173	1.4646	.0288	.0046	.5486	+242
Cascarilla	.8467	172	1.4652	.0305	.0049	.5494	0
Elder	.8468	172	1.4631	.0269	.0047	.5468	+ 15
Bay	.8508	171	1.4542	.0260	.0047	.5338	— 22
Gaultherilene	.8510	168	1.4614	.0271	.0049	.5422	
Nutmeg	.8518	167	1.4630	.0284	.0047	.5435	+ 49
" Penang	.8527	166	1.4634	.0274	.0049	.5434	+ 4
Carvene	.8530	166	1.4610	.0261	.0048	.5440	— 20
" Hamburg 2d dist.	.8545		1.4641	.0263	.0048	.5431	+ 86
Wormwood	.8565	160	1.4590	.0253	.0047	.5359	+ 46
Terebene	.8583	160	1.4670	.0275	.0048	.5440	0
Anise	.8580	160	1.4607	.0268	.0047	.5368	
Mint	.8600	160	1.4622	.0255	.0048	.5374	+ 30
Peppermint	.8602	175	1.4577	.0267	.0047	.5321	— 60
Laurel Turpentine	.8618	160	1.4637	.0260	.0047	.5380	+ 94
Thyme	.8635	160	1.4617	.0282	.0048	.5346	— 75
Turpentine I.	.8644	160	1.4612	.0250	.0047	.5335	+ 48
" II.	.8555	160	1.4590	.0256	.0047	.5365	— 87
" III.	.8614	160	1.4621	.0249		.5364	— 90
" IV	.8600	160	1.4613	.0254	.0047	.5364	— 88
Eucalyptus amygdalina	.8642	171	1.4696	.0323	.0049	.5434	—142

Source of Hydrocarbon.	Sp. Gr. at 20° C. = D	Boiling Point.	Index of Refraction of A at 20° C. μ A =	Dispersion, A to H	Δ 10°	μ A − 1 / D	Rotation.
Myrtle	.8690	163°	1.4565	.0248	.0047	.5253	+ 64°
Parsley	.8732	160	1.4665	.0291	.0046	.5355	— 44
Rosemary	.8805	163	1.4583	.0241	.0046	.5205	+ 8
Cloves	.9041	249	1.4898	.0284	.0045	.5417	
Rosewood	.9042	249	1.4878	.0277	.0045	.5395	— 11
Cubebs	.9062	260	1.4950	.0302	.0041	.5462	+ 59
Calamus	.9180	260	1.4930	.0322	.0042	.5370	+ 55
" Hamburg·	.9275	260	1.4976	.0337	.0043	.5365	+ 22
Cascarilla	.9212	254	1.4926	.0307	.0042	.5347	+ 72
Patchouli	.9211	254	1.4966	.0274	.0042	.5391	
" Penang	.9278	257	1.4963	.0275	.0044	.5349	— 90
" French	.9255	260	1.5009	.0262	.0042	.5412	
Colophene	.9391	315	1.5084	.0309	.0041	.5413	0

MISCELLANEOUS.

SOLUBILITIES OF SOME OF THE MORE COMMON PRECIPITATES.

FRESENIUS. — Quan. Analysis, pp. 740-749.

Bichloride of Potassium and Platinum. — In absence of free Hydrochloric acid one part of the salt dissolves in 12083 parts of alcohol of 97.5 per cent, in 3775, of 76 per cent, and in 1053, of 55 per cent. In presence of free Hydrochloric acid, in 1835, of 76 per cent.

Bichloride of Ammonium and Platinum. — In absence of free Hydrochloric acid, one part of the salt dissolves in 26535 parts of alcohol of 97.5 per cent, in 1406 parts of alcohol of 76 per cent, in 665 parts of alcohol of 55 per cent. In presence of free Hydrochloric acid, one part of the salt requires 672 parts of spirit of 76 per cent.

Carbonate of Barium. — One part dissolves in 14137 parts of water at 15–20°. One part dissolves in 15432 parts of boiling water. One part dissolves in 141000 parts of water containing Ammonia, Carbonate of Ammonium, and Chloride of Ammonium.

Fluosilicate of Barium. — One part of the salt requires 3802 parts of cold, or 3392 parts of boiling water for solution. One part of the salt requires from 640 to 733 parts of water acidified with Hydrochloric for solution.

Sulphate of Strontium. — One part of the salt dissolves in 6895 parts of water at 14°, in 9638 parts of boiling water, in 11862 parts of water containing Sulphuric and Hydrochloric acids, in 432 parts of Nitric acid of 4.8 per cent, in 474 parts of Hydrochloric of 8.5 per cent, and in 7843 parts of Acetic acid of 15.6 per cent.

Carbonate of Strontium. — One part of the salt requires 18045 parts of cold water for solution, and 56545 parts of a solution containing Carbonate of Ammonium and free Ammonia.

Carbonate of Calcium. — One part dissolves in 8834 parts of boiling, or in 10601 parts of cold water. One part requires 65246 parts of a solution of Carbonate of Ammonium and free Ammonia to dissolve it.

Basic Phosphate of Magnesium and Ammonium. — One part dissolves in 15293 parts of pure water, in 44600 parts of water containing free Ammonia, in 7548 parts of water containing Chloride of Ammonium, and in 15627 parts of water containing Chloride of Ammonium and Ammonia.

Magnesia. — One part dissolves in 55368 parts of water.

Basic Carbonate of Zinc. — One part dissolves in 44642 parts of water.

Carbonate of Lead. — One part requires 50551 parts of water, or 23450 parts of a dilute solution of Acetate of Ammonium, Carbonate of Ammonium, and Ammonia for Solution.

Sulphate of Lead. — One part dissolves in 22816 parts of pure water, and in 36504 parts of water containing Sulphuric acid.

WEIGHTS AND MEASURES.

Measure of Length.

One Myriametre	=	10000	Metres.
" Kilometre	=	1000	"
" Hectometre	=	100	"
" Decametre	=	10	"
" Metre	=	1	"
" Decimetre	=	0.1	"
" Centimetre	=	0.01	"
" Millimetre	=	0.001	"

Square Measure.

One Centiare	=		1 Square Metre.	
" Are	=	100	"	Metres.
" Hectare	=	10,000	"	"

Cubic Measure.

One Cubic Metre	=	1	Cubic Metre.	
" " Decimetre	=	0.001	"	
" " Centimetre	=	0.000,001	"	
" " Millimetre	=	0.000,000,001	"	
One Litre	=	1000 c.c.		

Weights.

One Kilogramme	=	1000	Grammes.
" Hectogramme	=	100	"
" Decagramme	=	10	"
" Gramme	=	1	"
" Decigramme	=	0.1	"
" Centigramme	=	0.01	"
" Milligramme	=	0.001	"

One Gramme = 15.4336 Grains = one Cubic Centimetre of water at 3.9° C. and 760 m.m. pressure.

VALUES OF ENGLISH WEIGHTS AND MEASURES IN FRENCH OR DECIMAL.

Wine Measure.

One Gallon	=	4.5435 Litres.
" Quart	=	1.1359 "
" Pint	=	.5679 "
" Gill	=	.1420 "
" Fluid Ounce	=	28.4 Cubic Centimetres.

Long Measure.

One Mile	=	1609.4083 Metres.
" Furlong	=	201.1760 "
" Rod	=	5.0297 "
" Yard	=	.9144 "
" Foot	=	.3048 "
" Inch	=	2.54 Centimetres.
" Line	=	.21 "

Square Measure.

One Square Mile	=	258.9894 Hectares.
" " Acre	=	40.4671 Ares.
" " Rood	=	10.1168 "
" " Rod	=	25.292 Square Metres.
" " Yard	=	0.8360 " "
" " Foot	=	.0929 " "
" " Inch	=	6.45 " Centimetres.

Cubic Measure.

One Cubic Foot	=	28315.3119 Cubic Centimetres.
" " Inch	=	16.3862 " "

Avoirdupois Weight.

One Ton, 2000 lbs.,	=	907.18530 Kilogrammes.
" Hundred Wt., 100 lbs.,	=	45.359265 "
" Pound	=	.45359265 "
" Ounce	=	28.3494 Grammes.
" Dram	=	1.7718 "

Troy Weight.

One Pound	=	373.242 Grammes.
" Ounce	=	31.1035 "
" Pennyweight	=	1.5522 "
" Grain	=	.0648 "

Apothecaries' Weight.

One Pound	=	373.242 Grammes.
" Ounce	=	31.1035 "
" Dram	=	3.8779 "
Scruple	=	1.296 "
" Grain	=	.0648 "

TABLE OF LOGARITHMS.

TABLE

1	0.00000	26	1.41497	51	1.70757	76	1.88081
2	0.30103	27	1.43136	52	1.71600	77	1.88649
3	0.47712	28	1.44716	53	1.72428	78	1.89209
4	0.60206	29	1.46240	54	1.73239	79	1.89763
5	0.69897	30	1.47712	55	1.74036	80	1.90309
6	0.77815	31	1.49136	56	1.74819	81	1.90849
7	0.84510	32	1.50515	57	1.75587	82	1.91381
8	0.90309	33	1.51851	58	1.76343	83	1.91908
9	0.95424	34	1.53148	59	1.77085	84	1.92428
10	1.00000	35	1.54407	60	1.77815	85	1.92942
11	1.04139	36	1.55630	61	1.78533	86	1.93450
12	1.07918	37	1.56820	62	1.79239	87	1.93952
13	1.11394	38	1.57978	63	1.79934	88	1.94448
14	1.14613	39	1.59106	64	1.80618	89	1.94939
15	1.17609	40	1.60206	65	1.81291	90	1.95424
16	1.20412	41	1.61278	66	1.81954	91	1.95904
17	1.23045	42	1.62325	67	1.82607	92	1.96379
18	1.25527	43	1.63347	68	1.83251	93	1.96848
19	1.27875	44	1.64345	69	1.83885	94	1.97313
20	1.30103	45	1.65321	70	1.84510	95	1.97772
21	1.32222	46	1.66276	71	1.85126	96	1.98227
22	1.34242	47	1.67210	72	1.85733	97	1.98677
23	1.36173	48	1.68124	73	1.86332	98	1.99123
24	1.38021	49	1.69020	74	1.86923	99	1.99564
25	1.39794	50	1.69897	75	1.87506	100	2.00000

NOTE. — In the following table the two leading figures in the first column of Logarithms are to be prefixed to all the numbers of the same horizontal line in the next nine columns, but when an asterisk (*) occurs, the two leading figures are to be taken from the next lower line.

	0	1	2	3	4	5	6	7	8	9
100	00000	043	087	130	173	217	260	303	346	389
101	432	475	518	561	604	647	689	732	775	817
102	860	903	945	988	*030	*072	*115	*157	*199	*242
103	01284	326	368	410	452	494	536	578	620	662
104	703	745	787	828	870	912	953	995	*036	*078
105	02119	160	202	243	284	325	366	407	449	490
106	531	572	612	653	694	735	776	816	857	898
107	938	979	*019	*060	*100	*141	*181	*222	*262	*302
108	03342	383	423	463	503	543	583	623	663	703
109	743	782	822	862	902	941	981	*021	*060	*100
110	04139	179	218	258	297	336	376	415	454	493
111	532	571	610	650	689	727	766	805	844	883
112	922	961	999	*038	*077	*115	*154	*192	*231	*269
113	05308	346	385	423	461	500	538	576	614	652
114	690	729	767	805	843	881	918	956	994	*032
115	06070	108	145	183	221	258	296	333	371	408
116	446	483	521	558	595	633	670	707	744	781
117	819	856	893	930	967	*004	*041	*078	*115	*151
118	07188	225	262	298	335	372	408	445	482	518
119	555	591	628	664	700	737	773	809	846	882
120	918	954	990	*027	*063	*099	*135	*171	*207	*243
121	08279	314	350	386	422	458	493	529	565	600
122	636	672	707	743	778	814	849	884	920	955
123	991	*026	*061	*096	*132	*167	*202	*237	*272	*307
124	09342	377	412	447	482	517	552	587	621	656
125	691	726	760	795	830	864	899	934	968	*003
126	10037	072	106	140	175	209	243	278	312	346
127	380	415	449	483	517	551	585	619	653	687
128	721	755	789	823	857	890	924	958	992	*025
129	11059	093	126	160	193	227	261	294	327	361
130	394	428	461	494	528	561	594	628	661	694
131	727	760	793	826	860	893	926	959	992	*024
132	12057	090	123	156	189	222	254	287	320	352
133	385	418	450	483	516	548	581	613	646	678
134	710	743	775	808	840	872	905	937	969	*001
135	13033	066	098	130	162	194	226	258	290	322
136	354	386	418	450	481	513	545	577	609	640
137	672	704	735	767	799	830	862	893	925	956
138	988	*019	*051	*082	*114	*145	*176	*208	*239	*270
139	14301	333	364	395	426	457	489	520	551	582

.	0	1	2	3	4	5	6	7	8	9
140	14613	644	675	706	737	768	799	829	860	891
141	922	953	983	*014	*045	*076	*106	*137	*168	*198
142	15229	259	290	320	351	381	412	442	473	503
143	534	564	594	625	655	685	715	746	776	806
144	836	866	897	927	957	987	*017	*047	*077	*107
145	16137	167	197	227	256	286	316	346	376	406
146	435	465	495	524	554	584	613	643	673	702
147	732	761	791	820	850	879	909	938	967	997
148	17026	056	085	114	143	173	202	231	260	289
149	319	348	377	406	435	464	493	522	551	580
150	609	638	667	696	725	754	782	811	840	869
151	898	926	955	984	*013	*041	*070	*099	*127	*156
152	18184	213	241	270	298	327	355	384	412	441
153	469	498	526	554	583	611	639	667	696	724
154	752	780	808	837	865	893	921	949	977	*005
155	19033	061	089	117	145	173	201	229	257	285
156	312	340	368	396	424	451	479	507	535	562
157	590	618	645	673	700	728	756	783	811	838
158	866	893	921	948	976	*003	*030	*058	*085	*112
159	20140	167	194	222	249	276	303	330	358	385
160	412	439	466	493	520	548	575	602	629	656
161	683	710	737	763	790	817	844	871	898	925
162	952	978	*005	*032	*059	*085	*112	*139	*165	*192
163	21219	245	272	299	325	352	378	405	431	458
164	484	511	537	564	590	617	643	669	696	722
165	748	775	801	827	854	880	906	932	958	985
166	22011	037	063	089	115	141	167	194	220	246
167	272	298	324	350	376	401	427	453	479	505
168	531	557	583	608	634	660	686	712	737	763
169	789	814	840	866	891	917	943	968	994	*019
170	23045	070	096	121	147	172	198	223	249	274
171	300	325	350	376	401	426	452	477	502	528
172	553	578	603	629	654	679	704	729	754	779
173	805	830	855	880	905	930	955	980	*005	*030
174	24055	080	105	130	155	180	204	229	254	279
175	304	329	353	378	403	428	452	477	502	527
176	551	576	601	625	650	674	699	724	748	773
177	797	822	846	871	895	920	944	969	993	*018
178	25042	066	091	115	139	164	188	212	237	261
179	285	310	334	358	382	406	431	455	479	503

.	0	1	2	3	4	5	6	7	8	9
180	25527	551	575	600	624	648	672	696	720	744
181	768	792	816	840	864	888	912	935	959	983
182	26007	031	055	079	102	126	150	174	198	221
183	245	269	293	316	340	364	387	411	435	458
184	482	505	529	553	576	600	623	647	670	694
185	717	741	764	788	811	834	858	881	905	928
186	951	975	998	*021	*045	*068	*091	*114	*138	*161
187	27184	207	231	254	277	300	323	346	370	393
188	416	439	462	485	508	531	554	577	600	623
189	646	669	692	715	738	761	784	807	830	852
190	875	898	921	944	967	989	*012	*035	*058	*081
191	28103	126	149	171	194	217	240	262	285	307
192	330	353	375	398	421	443	466	488	511	533
193	556	578	601	623	646	668	691	713	735	758
194	780	803	825	847	870	892	914	937	959	981
195	29003	026	048	070	092	115	137	159	181	203
196	226	248	270	292	314	336	358	380	403	425
197	447	469	491	513	535	557	579	601	623	645
198	667	688	710	732	754	776	798	820	842	863
199	885	907	929	951	973	994	*016	*038	*060	*081
200	30103	125	146	168	190	211	233	255	276	298
201	320	341	363	384	406	428	449	471	492	514
202	535	557	578	600	621	643	664	685	707	728
203	750	771	792	814	835	856	878	899	920	942
204	963	984	*006	*027	*048	*069	*091	*112	*133	*154
205	31175	197	218	239	260	281	302	323	345	366
206	387	408	429	450	471	492	513	534	555	576
207	597	618	639	660	681	702	723	744	765	785
208	806	827	848	869	890	911	931	952	973	994
209	32015	035	056	077	098	118	139	160	181	201
210	222	243	263	284	305	325	346	366	387	408
211	428	449	469	490	510	531	552	572	593	613
212	634	654	675	695	715	736	756	777	797	818
213	838	858	879	899	919	940	960	980	*001	*021
214	33041	062	082	102	122	143	163	183	203	224
215	244	264	284	304	325	345	365	385	405	425
216	445	465	486	506	526	546	566	586	606	626
217	646	666	686	706	726	746	766	786	806	826
218	846	866	885	905	925	945	965	985	*005	*025
219	34044	064	084	104	124	143	163	183	203	223

·	0	1	2	3	4	5	6	7	8	9
220	34242	262	282	301	321	341	361	380	400	420
221	439	459	479	498	518	537	557	577	596	616
222	635	655	674	694	713	733	753	772	792	811
223	830	850	869	889	908	928	947	967	986	*005
224	35025	044	064	083	102	122	141	160	180	199
225	218	238	257	276	295	315	334	353	372	392
226	411	430	449	468	488	507	526	545	564	583
227	603	622	641	660	679	698	717	736	755	774
228	793	813	832	851	870	889	908	927	946	965
229	984	*003	*021	*040	*059	*078	*097	*116	*135	*154
230	36173	192	211	229	248	267	286	305	324	342
231	361	380	399	418	436	455	474	493	511	530
232	549	568	586	605	624	642	661	680	698	717
233	736	754	773	791	810	829	847	866	884	903
234	922	940	959	977	996	*014	*033	*051	*070	*088
235	37107	125	144	162	181	199	218	236	254	273
236	291	310	328	346	365	383	401	420	438	457
237	475	493	511	530	548	566	585	603	621	639
238	658	676	694	712	731	749	767	785	803	822
239	840	858	876	894	912	931	949	967	985	*003
240	38021	039	057	075	093	112	130	148	166	184
241	202	220	238	256	274	292	310	328	346	364
242	382	399	417	435	453	471	489	507	525	543
243	561	578	596	614	632	650	668	686	703	721
244	739	757	775	792	810	828	846	863	881	899
245	917	934	952	970	987	*005	*023	*041	*058	*076
246	39094	111	129	146	164	182	199	217	235	252
247	270	287	305	322	340	358	375	393	410	428
248	445	463	480	498	515	533	550	568	585	602
249	620	637	655	672	690	707	724	742	759	777
250	794	811	829	846	863	881	898	915	933	950
251	967	985	*002	*019	*037	*054	*071	*088	*106	*123
252	40140	157	175	192	209	226	243	261	278	295
253	312	329	346	364	381	398	415	432	449	466
254	483	500	518	535	552	569	586	603	620	637
255	654	671	688	705	722	739	756	773	790	807
256	824	841	858	875	892	909	926	943	960	976
257	993	*010	*027	*044	*061	*078	*095	*111	*128	*145
258	41162	179	196	212	229	246	263	280	296	313
259	330	347	363	380	397	414	430	447	464	481

	0	1	2	3	4	5	6	7	8	9
260	41497	514	531	547	564	581	597	614	631	647
261	664	681	697	714	731	747	764	780	797	814
262	830	847	863	880	896	913	929	946	963	979
263	996	*012	*029	*045	*062	*078	*095	*111	*127	*144
264	42160	177	193	210	226	243	259	275	292	308
265	325	341	357	374	390	406	423	439	455	472
266	488	504	521	537	553	570	586	602	619	635
267	651	667	684	700	716	732	749	765	781	797
268	813	830	846	862	878	894	911	927	943	959
269	975	991	*008	*024	*040	*056	*072	*088	*104	*120
270	43136	152	169	185	201	217	233	249	265	281
271	297	313	329	345	361	377	393	409	425	441
272	457	473	489	505	521	537	553	569	584	600
273	616	632	648	664	680	696	712	727	743	759
274	775	791	807	823	838	854	870	886	902	917
275	933	949	965	981	996	*012	*028	*044	*059	*075
276	44091	107	122	138	154	170	185	201	217	232
277	248	264	279	295	311	326	342	358	373	389
278	404	420	436	451	467	483	498	514	529	545
279	560	576	592	607	623	638	654	669	685	700
280	716	731	747	762	778	793	809	824	840	855
281	871	886	902	917	932	948	963	979	994	*010
282	45025	040	056	071	086	102	117	133	148	163
283	179	194	209	225	240	255	271	286	301	317
284	332	347	362	378	393	408	423	439	454	469
285	484	500	515	530	545	561	576	591	606	621
286	637	652	667	682	697	712	728	743	758	773
287	788	803	818	834	849	864	879	894	909	924
288	939	954	969	984	*000	*015	*030	*045	*060	*075
289	46090	105	120	135	150	165	180	195	210	225
290	240	255	270	285	300	315	330	345	359	374
291	389	404	419	434	449	464	479	494	509	523
292	538	553	568	583	598	613	627	642	657	672
293	687	702	716	731	746	761	776	790	805	820
294	835	850	864	879	894	909	923	938	953	967
295	982	997	*012	*026	*041	*056	*070	*085	*100	*114
296	47129	144	159	173	188	202	217	232	246	261
297	276	290	305	319	334	349	363	378	392	407
298	422	436	451	465	480	494	509	524	538	553
299	567	582	596	611	625	640	654	669	683	698

	0	1	2	3	4	5	6	7	8	9
300	47712	727	741	756	770	784	799	813	828	842
301	857	871	885	900	914	929	943	958	972	986
302	48001	015	029	044	058	073	087	101	116	130
303	144	159	173	187	202	216	230	244	259	273
304	287	302	316	330	344	359	373	387	401	416
305	430	444	458	473	487	501	515	530	544	558
306	572	586	601	615	629	643	657	671	686	700
307	714	728	742	756	770	785	799	813	827	841
308	855	869	883	897	911	926	940	954	968	982
309	996	*010	*024	*038	*052	*066	*080	*094	*108	*122
310	49136	150	164	178	192	206	220	234	248	262
311	276	290	304	318	332	346	360	374	388	402
312	415	429	443	457	471	485	499	513	527	541
313	554	568	582	596	610	624	638	651	665	679
314	693	707	721	734	748	762	776	790	803	817
315	831	845	859	872	886	900	914	927	941	955
316	969	982	996	*010	*024	*037	*051	*065	*079	*092
317	50106	120	133	147	161	174	188	202	215	229
318	243	256	270	284	297	311	325	338	352	365
319	379	393	406	420	433	447	461	474	488	501
320	515	529	542	556	569	583	596	610	623	637
321	651	664	678	691	705	718	732	745	759	772
322	786	799	813	826	840	853	866	880	893	907
323	920	934	947	961	974	987	*001	*014	*028	*041
324	51055	068	081	095	108	121	135	148	162	175
325	188	202	215	228	242	255	268	282	295	308
326	322	335	348	362	375	388	402	415	428	441
327	455	468	481	495	508	521	534	548	561	574
328	587	601	614	627	640	654	667	680	693	706
329	720	733	746	759	772	786	799	812	825	838
330	851	865	878	891	904	917	930	943	957	970
331	983	996	*009	*022	*035	*048	*061	*075	*088	*101
332	52114	127	140	153	166	179	192	205	218	231
333	244	257	270	284	297	310	323	336	349	362
334	375	388	401	414	427	440	453	466	479	492
335	504	517	530	543	556	569	582	595	608	621
336	634	647	660	673	686	699	711	724	737	750
337	763	776	789	802	815	827	840	853	866	879
338	892	905	917	930	943	956	969	982	994	*007
339	53020	033	046	058	071	084	097	110	122	135

	0	1	2	3	4	5	6	7	8	9
340	53148	161	173	186	199	212	224	237	250	263
341	275	288	301	314	326	339	352	364	377	390
342	403	415	428	441	453	466	479	491	504	517
343	529	542	555	567	580	593	605	618	631	643
344	656	668	681	694	706	719	732	744	757	769
345	782	794	807	820	832	845	857	870	882	895
346	908	920	933	945	958	970	983	995	*008	*020
347	54033	045	058	070	083	095	108	120	133	145
348	158	170	183	195	208	220	233	245	258	270
349	283	295	307	320	332	345	357	370	382	394
350	407	419	432	444	456	469	481	494	506	518
351	531	543	555	568	580	593	605	617	630	642
352	654	667	679	691	704	716	728	741	753	765
353	777	790	802	814	827	839	851	864	876	888
354	900	913	925	937	949	962	974	986	998	*011
355	55023	035	047	060	072	084	096	108	121	133
356	145	157	169	182	194	206	218	230	242	255
357	267	279	291	303	315	328	340	352	364	376
358	388	400	413	425	437	449	461	473	485	497
359	509	522	534	546	558	570	582	594	606	618
360	630	642	654	666	678	691	703	715	727	739
361	751	763	775	787	799	811	823	835	847	859
362	871	883	895	907	919	931	943	955	967	979
363	991	*003	*015	*027	*038	*050	*062	*074	*086	*098
364	56110	122	134	146	158	170	182	194	205	217
365	229	241	253	265	277	289	301	312	324	336
366	348	360	372	384	396	407	419	431	443	455
367	467	478	490	502	514	526	538	549	561	573
368	585	597	608	620	632	644	656	667	679	691
369	703	714	726	738	750	761	773	785	797	808
370	820	832	844	855	867	879	891	902	914	926
371	937	949	961	972	984	996	*008	*019	*031	*043
372	57054	066	078	089	101	113	124	136	148	159
373	171	183	194	206	217	229	241	252	264	276
374	287	299	310	322	334	345	357	368	380	392
375	403	415	426	438	449	461	473	484	496	507
376	519	530	542	553	565	576	588	600	611	623
377	634	646	657	669	680	692	703	715	726	738
378	749	761	772	784	795	807	818	830	841	852
379	864	875	887	898	910	921	933	944	955	967

	0	1	2	3	4	5	6	7	8	9
380	57978	990	*001	*013	*024	*035	*047	*058	*070	*081
381	58092	104	115	127	138	149	161	172	184	195
382	206	218	229	240	252	263	274	286	297	309
383	320	331	343	354	365	377	388	399	410	422
384	433	444	456	467	478	490	501	512	524	535
385	546	557	569	580	591	602	614	625	636	647
386	659	670	681	692	704	715	726	737	749	760
387	771	782	794	805	816	827	838	850	861	872
388	883	894	906	917	928	939	950	961	973	984
389	995	*006	*017	*028	*040	*051	*062	*073	*084	*095
390	59106	118	129	140	151	162	173	184	195	207
391	218	229	240	251	262	273	284	295	306	318
392	329	340	351	362	373	384	395	406	417	428
393	439	450	461	472	483	494	506	517	528	539
394	550	561	572	583	594	605	616	627	638	649
395	660	671	682	693	704	715	726	737	748	759
396	770	780	791	802	813	824	835	846	857	868
397	879	890	901	912	923	934	945	956	966	977
398	988	999	*010	*021	*032	*043	*054	*065	*076	*086
399	60097	108	119	130	141	152	163	173	184	195
400	206	217	228	239	249	260	271	282	293	304
401	314	325	336	347	358	369	379	390	401	412
402	423	433	444	455	466	477	487	498	509	520
403	531	541	552	563	574	584	595	606	617	627
404	638	649	660	670	681	692	703	713	724	735
405	746	756	767	778	788	799	810	821	831	842
406	853	863	874	885	895	906	917	927	938	949
407	959	970	981	991	*002	*013	*023	*034	*045	*055
408	61066	077	087	098	109	119	130	140	151	162
409	172	183	194	204	215	225	236	247	257	268
410	278	289	300	310	321	331	342	352	363	374
411	384	395	405	416	426	437	448	458	469	479
412	490	500	511	521	532	542	553	563	574	584
413	595	606	616	627	637	648	658	669	679	690
414	700	711	721	731	742	752	763	773	784	794
415	805	815	826	836	847	857	868	878	888	899
416	909	920	930	941	951	962	972	982	993	*003
417	62014	024	034	045	055	066	076	086	097	107
418	118	128	138	149	159	170	180	190	201	211
419	221	232	242	252	263	273	284	294	304	315

	0	1	2	3	4	5	6	7	8	9
420	62325	335	346	356	366	377	387	397	408	418
421	428	439	449	459	469	480	490	500	511	521
422	531	542	552	562	572	583	593	603	613	624
423	634	644	655	665	675	685	696	706	716	726
424	737	747	757	767	778	788	798	808	818	829
425	839	849	859	870	880	890	900	910	921	931
426	941	951	961	972	982	992	*002	*012	*022	*033
427	63043	053	063	073	083	094	104	114	124	134
428	144	155	165	175	185	195	205	215	225	236
429	246	256	266	276	286	296	306	317	327	337
430	347	357	367	377	387	397	407	417	428	438
431	448	458	468	478	488	498	508	518	528	538
432	548	558	568	579	589	599	609	619	629	639
433	649	659	669	679	689	699	709	719	729	739
434	749	759	769	779	789	799	809	819	829	839
435	849	859	869	879	889	899	909	919	929	939
436	949	959	969	979	988	998	*008	*018	*028	*038
437	64048	058	068	078	088	098	108	118	128	137
438	147	157	167	177	187	197	207	217	227	237
439	246	256	266	276	286	296	306	316	326	335
440	345	355	365	375	385	395	404	414	424	434
441	444	454	464	473	483	493	503	513	523	532
442	542	552	562	572	582	591	601	611	621	631
443	640	650	660	670	680	689	699	709	719	729
444	738	748	758	768	777	787	797	807	816	826
445	836	846	856	865	875	885	895	904	914	924
446	933	943	953	963	972	982	992	*002	*011	*021
447	65031	040	050	060	070	079	089	099	108	118
448	128	137	147	157	167	176	186	196	205	215
449	225	234	244	254	263	273	283	292	302	312
450	321	331	341	350	360	369	379	389	398	408
451	418	427	437	447	456	466	475	485	495	504
452	514	523	533	543	552	562	571	581	591	600
453	610	619	629	639	648	658	667	677	686	696
454	706	715	725	734	744	753	763	772	782	792
455	801	811	820	830	839	849	858	868	877	887
456	896	906	916	925	935	944	954	963	973	982
457	992	*001	*011	*020	*030	*039	*049	*058	*068	*077
458	66087	096	106	115	124	134	143	153	162	172
459	181	191	200	210	219	229	238	247	257	266

	0	1	2	3	4	5	6	7	8	9
460	66276	285	295	304	314	323	332	342	351	361
461	370	380	389	398	403	417	427	436	445	455
462	464	474	483	492	502	511	521	530	539	549
463	558	567	577	586	596	605	614	624	633	642
464	652	661	671	680	689	699	708	717	727	736
465	745	755	764	773	783	792	801	811	820	829
466	839	848	857	867	876	885	894	904	913	922
467	932	941	950	960	969	978	987	997	*006	*015
468	67025	034	043	052	062	071	080	089	099	108
469	117	127	136	145	154	164	173	182	191	201
470	210	219	228	237	247	256	265	274	284	293
471	302	311	321	330	339	348	357	367	376	385
472	394	403	413	422	431	440	449	459	468	477
473	486	495	504	514	523	532	541	550	560	569
474	578	587	596	605	614	624	633	642	651	660
475	669	679	688	697	706	715	724	733	742	752
476	761	770	779	788	797	806	815	825	834	843
477	852	861	870	879	888	897	906	916	925	934
478	943	952	961	970	979	988	997	*006	*015	*024
479	68034	043	052	061	070	079	088	097	106	115
480	124	133	142	151	160	169	178	187	196	205
481	215	224	233	242	251	260	269	278	287	296
482	305	314	323	332	341	350	359	368	377	386
483	395	404	413	422	431	440	449	458	467	476
484	485	494	502	511	520	529	538	547	556	565
485	574	583	592	601	610	619	628	637	646	655
486	664	673	681	690	699	708	717	726	735	744
487	753	762	771	780	789	797	806	815	824	833
488	842	851	860	869	878	886	895	904	913	922
489	931	940	949	958	966	975	984	993	*002	*011
490	69020	028	037	046	055	064	073	082	090	099
491	108	117	126	135	144	152	161	170	179	188
492	197	205	214	223	232	241	249	258	267	276
493	285	294	302	311	320	329	338	346	355	364
494	373	381	390	399	408	417	425	434	443	452
495	461	469	478	487	496	504	513	522	531	539
496	548	557	566	574	583	592	601	609	618	627
497	636	644	653	662	671	679	688	697	705	714
498	723	732	740	749	758	767	775	784	793	801
499	810	819	827	836	845	854	862	871	880	888

	0	1	2	3	4	5	6	7	8	9
500	69897	906	914	923	932	940	949	958	966	975
501	984	992	*001	*010	*018	*027	*036	*044	*053	*062
502	70070	079	088	096	105	114	122	131	140	148
503	157	165	174	183	191	200	209	217	226	234
504	243	252	260	269	278	286	295	303	312	321
505	329	338	346	355	364	372	381	389	398	406
506	415	424	432	441	449	458	467	475	484	492
507	501	509	518	526	535	544	552	561	569	578
508	586	595	603	612	621	629	638	646	655	663
509	672	680	689	697	706	714	723	731	740	749
510	757	766	774	783	791	800	808	817	825	834
511	842	851	859	868	876	885	893	902	910	919
512	927	935	944	952	961	969	978	986	995	*003
513	71012	020	029	037	046	054	063	071	079	088
514	096	105	113	122	130	139	147	155	164	172
515	181	189	198	206	214	223	231	240	248	257
516	265	273	282	290	299	307	315	324	332	341
517	349	357	366	374	383	391	399	408	416	425
518	433	541	450	458	466	475	483	492	500	508
519	517	525	533	542	550	559	567	575	584	592
520	600	609	617	625	634	642	650	659	667	675
521	684	692	700	709	717	725	734	742	750	759
522	767	775	784	792	800	809	817	825	834	842
523	850	858	867	875	883	892	900	908	917	925
524	933	941	950	958	966	975	983	991	999	*008
525	72016	024	032	041	049	057	066	074	082	090
526	099	107	115	123	132	140	148	156	165	173
527	181	189	198	206	214	222	230	239	247	255
528	263	272	280	288	296	304	313	321	329	337
529	346	354	362	370	378	387	395	403	411	419
530	428	436	444	452	460	469	477	485	493	501
531	509	518	526	534	542	550	558	567	575	583
532	591	599	607	616	624	632	640	648	656	665
533	673	681	689	697	705	713	722	730	738	746
534	754	762	770	779	787	795	803	811	819	827
535	835	843	852	860	868	876	884	892	900	908
536	916	925	933	941	949	957	965	973	981	989
537	997	*006	*014	*022	*030	*038	*046	*054	*062	*070
538	73078	086	094	102	111	119	127	135	143	151
539	159	167	175	183	191	199	207	215	223	231

	0	1	2	3	4	5	6	7	8	9
540	73239	247	255	263	272	280	288	296	304	312
541	320	328	336	344	352	360	368	376	384	392
542	400	408	416	424	432	440	448	456	464	472
543	480	488	496	504	512	520	528	536	544	552
544	560	568	576	584	592	600	608	616	624	632
545	640	648	656	664	672	679	687	695	703	711
546	719	727	735	743	751	759	767	775	783	791
547	799	807	815	823	830	838	846	854	862	870
548	878	886	894	902	910	918	926	933	941	949
549	957	965	973	981	989	997	*005	*013	*020	*028
550	74036	044	052	060	068	076	084	092	099	107
551	115	123	131	139	147	155	162	170	178	186
552	194	202	210	218	225	233	241	249	257	265
553	273	280	288	296	304	312	320	327	335	343
554	351	359	367	374	382	390	398	406	414	421
555	429	437	445	453	461	468	476	484	492	500
556	507	515	523	531	539	547	554	562	570	578
557	586	593	601	609	617	624	632	640	648	656
558	663	671	679	687	695	702	710	718	726	733
559	741	749	757	764	772	780	788	796	803	811
560	819	827	834	842	850	858	865	873	881	889
561	896	904	912	920	927	935	943	950	958	966
562	974	981	989	997	*005	*012	*020	*028	*035	*043
563	75051	059	066	074	082	089	097	105	113	120
564	128	136	143	151	159	166	174	182	189	197
565	205	213	220	228	236	243	251	259	266	274
566	282	289	297	305	312	320	328	335	343	351
567	358	366	374	381	389	397	404	412	420	427
568	435	442	450	458	465	473	481	488	496	504
569	511	519	526	534	542	549	557	565	572	580
570	587	595	603	610	618	626	633	641	648	656
571	664	671	679	686	694	702	709	717	724	732
572	740	747	755	762	770	778	785	793	800	808
573	815	823	831	838	846	853	861	868	876	884
574	891	899	906	914	921	929	937	944	952	959
575	967	974	982	989	997	*005	*012	*020	*027	*035
576	76042	050	057	065	072	080	087	095	103	110
577	118	125	133	140	148	155	163	170	178	185
578	193	200	208	215	223	230	238	245	253	260
579	268	275	283	290	298	305	313	320	328	335

	0	1	2	3	4	5	6	7	8	9
580	76343	350	358	365	373	380	388	395	403	410
581	418	425	433	440	448	455	462	470	477	485
582	492	500	507	515	522	530	537	545	552	559
583	567	574	582	589	597	604	612	619	626	634
584	641	649	656	664	671	678	686	693	701	708
585	716	723	730	738	745	753	760	768	775	782
586	790	797	805	812	819	827	834	842	849	856
587	864	871	879	886	893	901	908	916	923	930
588	938	945	953	960	967	975	982	989	997	*004
589	77012	019	026	034	041	048	056	063	070	078
590	085	093	100	107	115	122	129	137	144	151
591	159	166	173	181	188	195	203	210	217	225
592	232	240	247	254	262	269	276	283	291	298
593	305	313	320	327	335	342	349	357	364	371
594	379	386	393	401	408	415	422	430	437	444
595	452	459	466	474	481	488	495	503	510	517
596	525	532	539	546	554	561	568	576	583	590
597	597	605	612	619	627	634	641	648	656	663
598	670	677	685	692	699	706	714	721	728	735
599	743	750	757	764	772	779	786	793	801	808
600	815	822	830	837	844	851	859	866	873	880
601	887	895	902	909	916	924	931	938	945	952
602	960	967	974	981	988	996	*003	*010	*017	*025
603	78032	039	046	053	061	068	075	082	089	097
604	104	111	118	125	132	140	147	154	161	168
605	176	183	190	197	204	211	219	226	233	240
606	247	254	262	269	276	283	290	297	305	312
607	319	326	333	340	347	355	362	369	376	383
608	390	398	405	412	419	426	433	440	447	455
609	462	469	476	483	490	497	504	512	519	526
610	533	540	547	554	561	569	576	583	590	597
611	604	611	618	625	633	640	647	654	661	668
612	675	682	689	696	704	711	718	725	732	739
613	746	753	760	767	774	781	789	796	803	810
614	817	824	831	838	845	852	859	866	873	880
615	888	895	902	909	916	923	930	937	944	951
616	958	965	972	979	986	993	*000	*007	*014	*021
617	79029	036	043	050	057	064	071	078	085	092
618	099	106	113	120	127	134	141	148	155	162
619	169	176	183	190	197	204	211	218	225	232

	0	1	2	.3	4	5	6	7	8	9
620	79239	246	253	260	267	274	281	288	295	302
621	309	316	323	330	337	344	351	358	365	372
622	379	386	393	400	407	414	421	428	435	442
623	449	456	463	470	477	484	491	498	505	511
624	518	525	532	539	546	553	560	567	574	581
625	588	595	602	609	616	623	630	637	644	650
626	657	664	671	678	685	692	699	706	713	720
627	727	734	741	748	754	761	768	775	782	789
628	796	803	810	817	824	831	837	844	851	858
629	865	872	879	886	893	900	906	913	920	927
630	934	941	948	955	962	969	975	982	989	996
631	80003	010	017	024	030	037	044	051	058	065
632	072	079	085	092	099	106	113	120	127	134
633	140	147	154	161	168	175	182	188	195	202
634	209	216	223	229	236	243	250	257	264	271
635	277	284	291	298	305	312	318	325	332	339
·636	346	353	359	366	373	380	387	393	400	407
637	414	421	428	434	441	448	455	462	468	475
638	482	489	496	502	509	516	523	530	536	543
639	550	557	564	570	577	584	591	598	604	611
640	618	625	632	638	645	652	659	665	672	679
641	686	693	699	706	713	720	726	733	740	747
642	754	760	767	774	781	787	794	801	808	814
643	821	828	835	841	848	855	862	868	875	882
644	889	895	902	909	916	922	929	936	943	949
645	956	963	969	976	983	990	996	*003	*010	*017
646	81023	030	037	043	050	057	064	070	077	084
647	090	097	104	111	117	124	131	137	144	151
648	158	164	171	178	184	191	198	204	211	218
649	224	231	238	245	251	258	265	271	278	285
650	291	298	305	311	318	325	331	338	345	351
651	358	365	371	378	385	391	398	405	411	418
652	425	431	438	445	451	458	465	471	478	485
653	491	498	505	511	518	525	531	538	544	551
654	558	564	571	578	584	591	598	604	611	617
655	624	631	637	644	651	657	664	671	677	684
656	690	697	704	710	717	723	730	737	743	750
657	757	763	770	776	783	790	796	803	809	816
658	823	829	836	842	849	856	862	869	875	882
659	889	895	902	908	915	921	928	935	941	948

LOGARITHMS.

	0	1	2	3	4	5	6	7	8	9
660	81954	961	968	974	981	987	994	*00	403	410
661	82020	027	033	040	046	053	060	066	477	485
662	086	092	099	105	112	119	125	13:	552	559
663	151	158	164	171	178	184	191	197	20:	634
664	217	223	230	236	243	249	256	263		708
665	282	289	295	302	308	315	321	328	775	782
666	347	354	360	367	373	380	387	393	849	856
667	413	419	426	432	439	445	452	458	923	930
668	478	484	491	497	504	510	517	52	997	*004
669	543	549	556	562	569	575	582	58	070	078
670	607	614	620	627	633	640	646	65:	144	151
671	672	679	685	692	698	705	711	71:	217	225
672	737	743	750	756	763	769	776	78:	291	298
673	802	808	814	821	827	834	840	847	364	371
674	866	872	879	885	892	898	905	91:	437	444
675	930	937	943	950	956	963	969	975	982	988
676	995	*001	*008	*014	*020	*027	*033	*040	*046	*052
677	83059	065	072	078	085	091	097	104	110	117
678	123	129	136	142	149	155	161	168	174	181
679	187	193	200	206	213	219	225	23:	238	245
680	251	257	264	270	276	283	289	296	302	308
681	315	321	327	334	340	347	353	359	366	372
682	378	385	391	398	404	410	417	423	429	436
683	442	448	455	461	467	474	480	487	493	499
684	506	512	518	525	531	537	544	550	556	563
685	569	575	582	588	594	601	607	613	620	626
686	632	639	645	651	658	664	670	677	683	689
687	696	702	708	715	721	727	734	740	746	753
688	759	765	771	778	784	790	797	803	809	816
689	822	828	835	841	847	853	860	866	872	879
690	885	891	897	904	910	916	923	929	935	942
691	948	954	960	967	973	979	985	992	998	*004
692	84011	017	023	029	036	042	048	055	061	067
693	073	080	086	092	098	105	111	117	123	130
694	136	142	148	155	161	167	173	180	186	192
695	198	205	211	217	223	230	236	242	248	255
696	261	267	273	280	286	292	298	305	311	317
697	323	330	336	342	348	354	361	367	373	379
698	386	392	398	404	410	417	423	429	435	442
699	448	454	460	466	473	479	485	491	497	504

	0	1	2	3	4	5	6	7	8	9
620	79239	516	522	528	535	541	547	553	559	566
621		578	584	590	597	603	609	615	621	628
622		640	646	652	658	665	671	677	683	689
623		702	708	714	720	726	733	739	745	751
624	757	763	770	776	782	788	794	800	807	813
625		825	831	837	844	850	856	862	868	874
626		887	893	899	905	911	917	924	930	936
627		948	954	960	967	973	979	985	991	997
628	790	009	016	022	028	034	040	046	052	058
629	865	071	077	083	089	095	101	107	114	120
630	933	132	138	144	150	156	163	169	175	181
631	80003	193	199	205	211	217	224	230	236	242
632		254	260	266	272	278	285	291	297	303
633		315	321	327	333	339	345	352	358	364
634		376	382	388	394	400	406	412	418	425
715	431	437	443	449	455	461	467	473	479	485
716	491	497	503	509	516	522	528	534	540	546
717	552	558	564	570	576	582	588	594	600	606
718	612	618	625	631	637	643	649	655	661	667
719	673	679	685	691	697	703	709	715	721	727
720	733	739	745	751	757	763	769	775	781	788
721	794	800	806	812	818	824	830	836	842	848
722	854	860	866	872	878	884	890	896	902	908
723	914	920	926	932	938	944	950	956	962	968
724	974	980	986	992	998	*004	*010	*016	*022	*028
725	86034	040	046	052	058	064	070	076	082	088
726	094	100	106	112	118	124	130	136	141	147
727	153	159	165	171	177	183	189	195	201	207
728	213	219	225	231	237	243	249	255	261	267
729	273	279	285	291	297	303	308	314	320	326
730	832	338	344	350	356	362	368	374	380	386
731	392	398	404	410	415	421	427	433	439	445
732	451	457	463	469	475	481	487	493	499	504
733	510	516	522	528	534	540	546	552	558	564
734	570	576	581	587	593	599	605	611	617	623
735	629	635	641	646	652	658	664	670	676	682
736	688	694	700	705	711	717	723	729	735	741
737	747	753	759	764	770	776	782	788	794	800
738	806	812	817	823	829	835	841	847	853	859
739	864	870	876	882	888	894	900	906	911	917

	0	1	2	3	4	5	6	7	8	9
740	86923	929	935	941	947	953	958	964	970	976
741	982	988	994	999	*005	*011	*017	*023	*029	*035
742	87040	046	052	058	064	070	075	081	087	093
743	099	105	111	116	122	128	134	140	146	151
744	157	163	169	175	181	186	192	198	204	210
745	216	221	227	233	239	245	251	256	262	268
746	274	280	286	291	297	303	309	315	320	326
747	332	338	344	349	355	361	367	373	379	384
748	390	396	402	408	413	419	425	431	437	442
749	448	454	460	466	471	477	483	489	495	500
750	506	512	518	523	529	535	541	547	552	558
751	564	570	576	581	587	593	599	604	610	616
752	622	628	633	639	645	651	656	662	668	674
753	679	685	691	697	703	708	714	720	726	731
754	737	743	749	754	760	766	772	777	783	789
755	795	800	806	812	818	823	829	835	841	846
756	852	858	864	869	875	881	887	992	898	904
757	910	915	921	927	933	938	944	950	955	961
758	967	973	978	984	990	996	*001	*007	*013	*018
759	88024	030	036	041	047	053	058	064	070	076
760	081	087	093	098	104	110	116	121	127	133
761	138	144	150	156	161	167	173	178	184	190
762	195	201	207	213	218	224	230	235	241	247
763	252	258	264	270	275	281	287	292	298	304
764	309	315	321	326	332	338	343	349	355	360
765	366	372	377	383	389	395	400	406	412	417
766	423	429	434	440	446	451	457	463	468	474
767	480	485	491	497	502	508	513	519	525	530
768	536	542	547	553	559	564	570	576	581	587
769	593	598	604	610	615	621	627	632	638	643
770	649	655	660	666	672	677	683	689	694	700
771	705	711	717	722	728	734	739	745	750	756
772	762	767	773	779	784	790	795	801	807	812
773	818	824	829	835	840	846	852	857	863	868
774	874	880	885	891	897	902	908	913	919	925
775	930	936	941	947	953	958	964	969	975	981
776	986	992	997	*003	*009	*014	*020	*025	*031	*037
777	89042	048	053	059	064	070	076	081	087	092
778	098	104	109	115	120	126	131	137	143	148
779	154	159	165	170	176	182	187	193	198	204

	0	1	2	3	4	5	6	7	8	9
780	89209	215	221	226	232	237	243	248	254	260
781	265	271	276	282	287	293	298	304	310	315
782	321	326	332	337	343	348	354	360	365	371
783	376	382	387	393	398	404	409	415	421	426
784	432	437	443	448	454	459	465	470	476	481
785	487	492	498	504	509	515	520	526	531	537
786	542	548	553	559	564	570	575	581	586	592
787	597	603	609	614	620	625	631	636	642	647
788	653	658	664	669	675	680	686	691	697	702
789	708	713	719	724	730	735	741	746	752	757
790	763	768	774	779	785	790	796	801	807	812
791	818	823	829	834	840	845	851	856	862	867
792	873	878	883	889	894	900	905	911	916	922
793	927	933	938	944	949	955	960	966	971	977
794	982	988	993	998	*004	*009	*015	*020	*026	*031
795	90037	042	048	053	059	064	069	075	080	086
796	091	097	102	108	113	119	124	129	135	140
797	146	151	157	162	168	173	179	184	189	195
798	200	206	211	217	222	227	233	238	244	249
799	255	260	266	271	276	282	287	293	298	304
800	309	314	320	325	331	336	342	347	352	358
801	363	369	374	380	385	390	396	401	407	412
802	417	423	428	434	339	445	450	455	461	466
803	472	477	482	488	493	499	504	509	515	520
804	526	531	536	542	547	553	558	563	569	574
805	580	585	590	596	601	607	612	617	623	628
806	634	639	644	650	655	660	666	671	677	682
807	687	693	698	703	709	714	720	725	730	736
808	741	747	752	757	763	768	773	779	784	789
809	795	800	806	811	816	822	827	832	838	843
810	849	854	859	865	870	875	881	886	891	897
811	902	907	913	818	924	929	934	940	945	950
812	956	961	966	972	977	982	988	993	998	*004
813	91009	014	020	025	030	036	041	046	052	057
814	062	068	073	078	084	089	094	100	105	110
815	116	121	126	132	137	142	148	153	158	164
816	169	174	180	185	190	196	201	206	212	217
817	222	228	233	238	243	249	254	259	265	270
818	275	281	386	291	297	302	307	312	318	323
819	328	334	339	344	350	355	360	365	371	376

	0	1	2	3	4	5	6	7	8	9
820	91381	387	392	397	403	408	413	418	424	429
821	434	440	445	450	455	461	466	471	477	482
822	487	492	498	503	508	514	519	524	529	535
823	540	545	551	556	561	566	572	577	582	587
824	593	598	603	609	614	619	624	630	635	640
825	645	651	656	661	666	672	677	682	687	693
826	698	703	709	714	719	724	730	735	740	745
827	751	756	761	766	772	777	782	787	793	798
828	803	808	814	819	824	829	834	840	845	850
829	855	861	866	871	876	882	887	892	897	903
830	908	913	918	924	929	934	939	944	950	955
831	960	965	971	976	981	986	991	997	*002	*007
832	92012	018	023	028	033	038	044	049	054	059.
833	065	070	075	080	085	091	096	101	106	111
834	117	122	127	132	137	143	148	153	158	163
835	169	174	179	184	189	195	200	205	210	215
836	221	226	231	236	241	247	252	257	262	267
837	273	278	283	288	293	298	304	309	314	319
838	324	330	335	340	345	350	355	361	366	371
839	376	381	387	392	397	402	407	412	418	423
840	428	433	438	443	449	454	459	464	469	474
841	480	485	490	495	500	505	511	516	521	526
842	531	536	542	547	552	557	562	567	572	578
843	583	588	593	598	603	609	614	619	624	629
844	634	639	645	650	655	660	665	670	675	681
845	686	691	696	701	706	711	716	722	727	732
846	737	742	747	752	758	763	768	773	778	783
847	788	793	799	804	809	814	819	824	829	834
848	840	845	850	855	860	865	870	875	881	886
849	891	896	901	906	911	916	921	927	932	937
850	942	947	952	957	962	967	973	978	983	988
851	993	998	*003	*008	*013	*018	*024	*029	*034	*039
852	93044	049	054	059	064	069	075	080	085	090
853	095	100	105	110	115	120	125	131	136	141
854	146	151	156	161	166	171	176	181	186	192
855	197	202	207	212	217	222	227	232	237	242
856	247	252	258	263	268	273	278	283	288	293
857	298	303	308	313	318	323	328	334	339	344
858	349	354	359	364	369	374	379	384	389	394
859	399	404	409	414	420	425	430	435	440	445

	0	**1**	**2**	**3**	**4**	**5**	**6**	**7**	**8**	**9**
860	93450	455	460	465	470	475	480	485	490	495
861	500	505	510	515	520	526	531	536	541	546
862	551	556	561	566	571	576	581	586	591	596
863	601	606	611	616	621	626	631	636	641	646
864	651	656	661	666	671	676	682	687	692	697
865	702	707	712	717	722	727	732	737	742	747
866	752	757	762	767	772	777	782	787	792	797
867	802	807	812	817	822	827	832	837	842	847
868	852	857	862	867	872	877	882	887	892	897
869	902	907	912	917	922	927	932	937	942	947
870	952	957	962	967	972	977	982	987	992	997
871	94002	007	012	017	022	027	032	037	042	047
872	052	057	062	067	072	077	082	086	091	096
873	101	106	111	116	121	126	131	136	141	146
874	151	156	161	166	171	176	181	186	191	196
875	201	206	211	216	221	226	231	236	240	245
876	250	255	260	265	270	275	280	285	290	295
877	300	305	310	315	320	325	330	335	340	345
878	349	354	359	364	369	374	379	384	389	394
879	399	404	409	414	419	424	429	433	438	443
880	448	453	458	463	468	473	478	483	488	493
881	498	503	507	512	517	522	527	532	537	542
882	547	552	557	562	567	571	576	581	586	591
883	596	601	606	611	616	721	626	630	635	640
884	645	650	655	660	665	670	675	680	685	689
885	694	699	704	709	714	719	724	729	734	738
886	743	748	753	758	763	768	773	778	783	787
887	792	797	802	807	812	817	822	827	832	836
888	841	846	851	856	861	866	871	876	880	885
889	890	895	900	905	910	915	919	924	929	934
890	939	944	949	954	959	963	968	973	978	983
891	988	993	998	*002	*007	*012	*017	*022	*027	*032
892	95036	041	046	051	056	061	066	071	075	080
893	085	090	095	100	105	109	114	119	124	129
894	134	139	143	148	153	158	163	168	173	177
895	182	187	192	197	202	207	211	216	221	226
896	231	236	240	245	250	255	260	265	270	274
897	279	284	289	294	299	303	308	313	318	423
898	328	332	337	342	347	352	357	361	366	371
899	376	381	386	390	395	400	405	410	415	419

	0	1	2	3	4	5	6	7	8	9
900	95424	429	434	439	444	448	453	458	463	468
901	472	477	482	487	492	497	501	506	511	516
902	521	525	530	535	540	545	550	554	559	564
903	569	574	578	583	588	593	598	602	607	612
904	617	622	626	631	636	641	646	650	655	660
905	665	670	674	679	684	689	694	698	703	708
906	713	718	722	727	732	737	742	746	751	756
907	761	766	770	775	780	785	789	794	799	804
908	809	813	818	823	828	832	837	842	847	852
909	856	861	866	871	875	880	885	890	895	899
910	904	909	914	918	923	928	933	938	942	947
911	952	957	961	966	971	976	980	985	990	995
912	999	*004	*009	*014	*019	*023	*028	*033	*038	*042
913	96047	052	057	061	066	071	076	080	085	090
914	095	099	104	109	114	118	123	128	133	137
915	142	147	152	156	161	166	171	175	180	185
916	190	194	199	204	209	213	218	223	227	232
917	237	242	246	251	256	261	265	270	275	280
918	284	289	294	298	303	308	313	317	322	327
919	332	336	341	346	350	355	360	365	369	374
920	379	384	388	393	398	402	407	412	417	421
921	426	431	435	440	445	450	454	459	464	468
922	473	478	483	487	492	497	501	506	511	515
923	520	525	530	534	539	544	548	553	558	562
924	567	572	577	581	586	591	595	600	605	609
925	614	619	624	628	633	638	642	647	652	656
926	661	666	670	675	680	685	689	694	699	703
927	708	713	717	722	727	731	736	741	745	750
928	755	759	764	769	774	778	783	788	792	797
929	802	806	811	816	820	825	830	834	839	844
930	848	853	858	862	867	872	876	881	886	890
931	895	900	904	909	914	918	923	928	932	937
932	942	946	951	956	960	965	970	974	979	984
933	988	993	997	*002	*007	*011	*016	*021	*025	*030
934	97035	039	044	049	053	058	063	067	072	077
935	081	086	090	095	100	104	109	114	118	123
936	128	132	137	142	146	151	155	160	165	169
937	174	179	183	188	192	197	202	206	211	216
938	220	225	230	234	239	243	248	253	257	262
939	267	271	276	280	285	290	294	299	304	308

	0	1	2	3	4	5	6	7	8	9
940	97313	317	322	327	331	336	340	345	350	354
941	359	364	368	373	377	382	387	391	396	400
942	405	410	414	419	424	428	433	437	442	447
943	451	456	460	465	470	474	479	483	488	493
944	497	502	506	511	516	520	525	529	534	539
945	543	548	552	557	562	566	571	575	580	585
946	589	594	598	603	607	612	617	621	626	630
947	635	640	644	649	653	658	663	667	672	676
948	681	685	690	695	699	704	708	713	717	722
949	727	731	736	740	745	749	754	759	763	768
950	772	777	782	786	791	795	800	804	809	813
951	818	823	827	832	836	841	845	850	855	859
952	864	868	873	877	882	886	891	896	900	905
953	909	914	918	923	928	932	937	941	946	950
954	955	959	964	968	973	978	982	987	991	996
955	98000	005	009	014	019	023	028	032	037	041
956	046	050	055	959	064	068	073	078	082	087
957	091	096	100	105	109	114	118	123	127	132
958	137	141	146	150	155	159	164	168	173	177
959	182	186	191	195	200	204	209	214	218	223
960	227	232	236	241	245	250	254	259	263	268
961	272	277	281	286	290	295	299	304	308	313
962	318	322	327	331	336	340	345	349	354	358
963	363	367	372	376	381	385	390	394	399	403
964	408	412	417	421	426	430	435	439	444	448
965	453	457	462	466	471	475	480	484	489	493
966	498	502	507	511	516	520	525	529	534	538
967	543	547	552	556	561	565	570	574	579	583
968	588	592	597	601	605	610	614	619	623	628
969	632	637	641	646	650	655	659	664	668	673
970	677	682	686	691	695	700	704	709	713	717
971	722	726	731	735	740	744	749	753	758	762
972	767	771	776	780	784	789	793	798	802	807
973	811	816	820	825	829	834	838	843	847	851
974	856	860	865	869	874	878	883	887	892	896
975	900	905	909	914	918	923	927	932	936	941
976	945	949	954	958	963	967	972	976	981	985
977	989	994	998	*003	*007	*012	*016	*021	*025	*029
978	99034	038	043	047	052	056	061	065	069	074
979	078	083	087	092	096	100	105	109	114	118

	0	1	2	3	4	5	6	7	8	9
980	99123	127	131	136	140	145	149	154	158	162
981	167	171	176	180	185	189	193	198	202	207
982	211	216	220	224	229	233	238	242	247	251
983	255	260	264	269	273	277	282	286	291	295
984	300	304	308	313	317	322	326	330	335	339
985	344	348	352	357	361	366	370	374	379	383
986	388	392	396	401	405	410	414	419	423	427
987	432	436	441	445	449	454	458	463	467	471
988	476	480	484	489	493	498	502	506	511	515
989	520	524	528	533	537	542	546	550	555	559
990	564	568	572	577	581	585	590	594	599	603
991	607	612	616	621	625	629	634	638	642	647
992	651	656	660	664	669	673	677	682	686	691
993	695	699	704	708	712	717	721	726	730	734
994	739	743	747	752	756	760	765	769	774	778
995	782	787	791	795	800	804	808	813	817	822
996	826	830	835	839	843	848	852	856	861	865
997	870	874	878	883	887	891	896	900	904	909
998	913	917	922	926	930	935	939	944	948	952
999	957	961	965	970	974	978	983	987	991	996

3157

QD65
S5

www.ingramcontent.com/pod-product-compliance
Lightning Source LLC
Chambersburg PA
CBHW021708210326
41599CB00013B/1575